62.2 User's Manual

ANSI/ASHRAE Standard 62.2-2010
Ventilation and Acceptable Indoor Air Quality in Low-Rise Residential Buildings

62.2 User's Manual

ANSI/ASHRAE Standard 62.2-2010
Ventilation and Acceptable Indoor Air Quality in Low-Rise Residential Buildings

® American Society of Heating, Refrigerating and Air-Conditioning Engineers, Inc.

ASHRAE Research: Improving the Quality of Life

The American Society of Heating, Refrigerating and Air-Conditioning Engineers (ASHRAE) is the world's foremost technical society in the fields of heating, ventilation, air conditioning and refrigeration. Its 55,000 members worldwide are individuals who share ideas, identify the need for and support research, and write the industry's standards for testing and practice. The result of these efforts is that engineers are better able to keep indoor environments safe and productive while protecting and preserving the outdoors for generations to come.

One of the ways that ASHRAE supports its members' and the industry's need for information is through ASHRAE Research. Thousands of individuals and companies support ASHRAE Research annually, enabling ASHRAE to report new data about material properties and building physics and to promote the application of innovative technologies.

ASHRAE Research contributed significantly to the material in this book.

For more information about ASHRAE Research or to become a member of ASHRAE, contact ASHRAE, 1791 Tullie Circle, N.E., Atlanta, GA 30329 USA; telephone +1-404-636-8400; www.ashrae.org.

1791 Tullie Circle
Atlanta, GA 30329

Printed in the United States of America.

ISBN 978-1-933742-99-1

Any updates/errata to this publication will be posted on the ASHRAE Web site at www.ashrae.org/publicationupdates.

Contents

Preface

This User's Manual contains explanatory material, examples and background material which are intended to aid the user in designing and constructing residential buildings which comply with ASHRAE Standard 62.2-2010, *Ventilation and Acceptable Indoor Air Quality in Low-Rise Residential Buildings.* Standard 62.2 was written in mandatory language in order to be code enforceable, except for three informative language appendices. This manual does not reproduce the requirements of the standard, but rather paraphrases and explains them. This manual is intended to be used in conjunction with the standard.

The User's Manual is intended to be used by:

- Residential Building Contractors
- Residential Building Subcontractors
- Architects
- Engineers

It may also be useful for:

- Code Officials
- Governmental Agencies
- Homeowners

Standard 62.2 is a dynamic document that is maintained and updated by ASHRAE under their "Continuous Maintenance" procedures. Periodically, ASHRAE will develop and publish addenda which modify or add to the standard. ASHRAE plans to publish a new version of the standard on a three year cycle, which will include all the addenda that have been approved since the previous publication. In between the published versions of the standard, ASHRAE also plans to publish a Supplement, which will contain any addenda approved in the first 18 months of the three year revision cycle.

Standard 62.2 was originally published in 2003, as Standard 62.2-2003. This User's Manual applies to the version of the standard published in 2010, which is known as 62.2-2010. When using this manual to comply with any codes based on Standard 62.2, check to confirm that the code is based on the 2010version. If it is based on some other version, those parts of this manual which apply to sections that are modified will not be reliable.

In addition to addenda, ASHRAE also periodically publishes official interpretations of its standards. Interpretations do not change the standard, but are used to clarify the meaning of existing requirements. Any person can submit a "Request for Interpretation" to ASHRAE. Interpretations are limited to the meaning of the standard, and may not be used for design reviews or requests for changes to the standard. Requests for Interpretation can be sent to:

Manager of Standards
ASHRAE
1791 Tullie Circle, NE
Atlanta, GA 30329-2305

When a copy of the standard is purchased from ASHRAE, any interpretations which apply to the standard are included in the package. The ASHRAE website also lists interpretations, and users can check there for any interpretations that may have been issued since they purchased the standard. (www.ASHRAE.org)

This User's Manual is organized to follow the organization of the standard. All the sections and subsections of the manual follow those of the standard. Each section of the manual contains explanatory material which is intended to assist the user in understanding the requirements of the standard and how to comply with them. Many sections also include examples which give specific means of complying with the standard. Some sections also include background material which explains why the standard includes particular requirements, how equipment tests are performed by the manufacturer or testing organizations, or theoretical information intended to provide a deeper understanding of the requirements and means of compliance.

Acknowledgements

The 2010 edition of the 62.2 User's Manual was based heavily on content originally developed for the 2004 version. The 2010 version was primarily developed by Roger Hedrick of Architectural Energy Corporation under contract to the American Society of Heating, Refrigerating and Air-Conditioning Engineers, Inc. (ASHRAE). Substantial contributions were also made by members of the Project Management Subcommittee chaired by Amy Musser, and including Rick Karg, Iain Walker, Eric Werling and Ted Williams, as well as the chair of SSPC 62.2, Steve Emmerich.

The 2004 edition of the User's Manual was written by Roger Hedrick, Terry Brennan, Don Stevens, and H. E. "Barney" Burroughs.

Section 1 – Purpose

This section of the standard does not contain any requirements. It does describe why the standard exists, and what compliance with the standard is expected to achieve.

The purpose is a single sentence, but it has two parts. The first part, "defines the roles of and minimum requirements for mechanical and natural ventilation systems and the building envelope" describes the parts of the building that the standard will primarily be addressing with its requirements: the ventilation system (mechanical or natural) and the building envelope. Don't interpret this as a statement of the scope of the standard, it includes requirements for additional parts of the building too. (See Section 2 – Scope, for more on this.)

The second part of the purpose, "intended to provide acceptable indoor air quality in low-rise residential buildings" describes why the requirements of the standard exist.

Acceptable indoor air quality is a term that is defined in Section 3 – Definitions, to mean air that is neither irritating nor unhealthy. Indoor air that is not acceptable is air that smells bad, contains irritating contaminants such as pollen or other allergens, or contains dangerous chemicals at concentrations that may cause health effects. Unacceptable indoor air can have one, two or all three of these characteristics. It may have odors only, but not be unhealthy, or it might be air that seems fine but contains dangerous concentrations of toxic chemicals that cannot be sensed by the occupants.

In summary, the purpose states that the standard prescribes ventilation, building envelope and other measures intended to provide residential indoor air quality that is safe and pleasant.

Section 2 – Scope

This section of the standard defines its scope in a number of different ways. First, the scope describes where the standard is intended to be applied, in terms of the types of buildings where it should be used. Second, the scope describes what the standard addresses, that is, what kinds of issues in the house that affect indoor air quality are covered. This includes a list of reasons why acceptable indoor air quality may not be achieved, even if all the requirements of the standard are met. Third, the scope lists equipment that is not addressed by the standard.

The standard applies to all residential spaces intended for human occupancy in single family houses or in multifamily buildings up to three stories. The phrase "intended for human occupancy" means that any space that people will normally enter is covered. This includes:

- living rooms,
- bedrooms,
- kitchens,
- bathrooms,
- hallways,
- closets,
- store rooms,
- laundries,
- garages, and
- basements

Other spaces within the building, such as attics or crawlspaces are not "covered" by the standard, in that they are not used in calculating ventilation rates, and they are not required to be ventilated to comply with the standard. The envelope requirements of the standard, however, still apply even if they surround these "non-covered" spaces.

This standard applies to spaces within single family buildings and multi-family buildings that are three stories or less above grade.

Single Family Dwellings

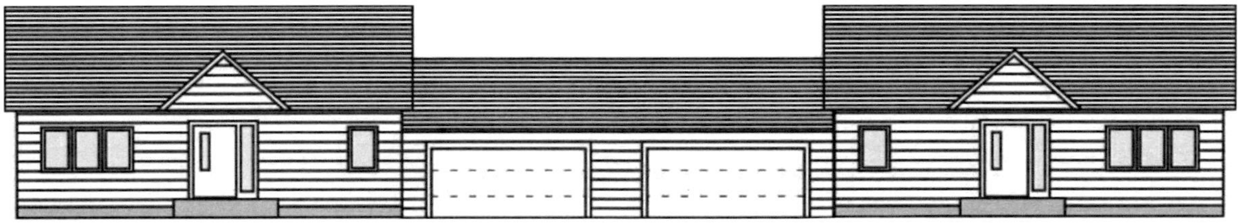

Duplex

Row Houses

Three Story Multi-Family

It does not apply to multi-family buildings greater than three stories, hotels, motels, dormitories, nursing homes, or prisons. In urban areas, a common building type has retail or other commercial space on the first floor, with residential apartments above. If the building is two or three stories, then this standard applies to the residential spaces, while the commercial space is covered by ASHRAE Standard 62.1-2010, *Ventilation and Indoor Air Quality*.

The second aspect of the scope is the types of issues that may occur or exist in a building that the standard addresses. The standard considers chemical, physical and biological air contaminants. It does not consider thermal comfort, which is addressed by ASHRAE Standard 55-2010, *Thermal Environmental Conditions for Human Occupancy*. It is intended to provide measures to address any type of indoor air contaminant which might be found in typical housing.

The standard provides requirements for whole-building ventilation and local exhaust ventilation, specifying minimum rates. These are engineering controls for contaminants released indoors. To make the performance of the ventilation systems more robust other requirements are included for:

- control strategies,
- sound levels,
- duct design,
- maximum flow rates for unbalanced systems,
- isolation of garages, and
- duct tightness.

ASHRAE defines these requirements as the minimum controls needed to accommodate ordinary residential air contaminant sources, including high humidity. Water vapor is not a contaminant in itself, but high humidity often results in mold problems, which can become an indoor air quality problem.

Meeting the requirements of this standard does not guarantee that air quality problems will not occur in a home. The standard lists a number of possible causes:

a) The requirements of the standard were developed to provide minimum requirements for typical housing and occupants. However, there will be a range of conditions in houses which may result in higher contaminant loads than typical, with the result that a ventilation system which complies with the standard may not be adequate to provide good IAQ at all times. There is also a wide range of susceptibility to various contaminants among occupants. A person who is particularly sensitive to a contaminant may perceive the air to be unacceptable despite the proper installation and operation of systems which comply with the standard.

b) Research has shown that occupants who are uncomfortable or under stress are extra susceptible to indoor air quality issues, or their discomfort may manifest itself in symptoms which are commonly associated with poor indoor air quality.

Four Story Multi-Family – Does Not Apply

c) When the outdoor air experiences high contamination levels, the ventilation system can introduce contaminated air that may result in poorer indoor air quality. A common example is when a house is located near a pollution source, such as a busy highway.

d) A building that meets this standard may have the ventilation equipment turned off, broken, or performing poorly due to lack of maintenance. The standard only requires that information on proper operation and maintenance of the system be provided to the owner or occupant. The occupant's use of that information is outside the scope of the standard.

e) On occasion, there may be releases of contaminants which are either very odorous, especially toxic, or just released in large amounts. Such cases may overwhelm ventilation systems designed in accordance with this standard. For example, certain hobby activities, heavy spraying of pesticides, food burned while cooking, indoor barbequing, operation of internal combustion equipment such as backup generators, etc. would all be considered "high-polluting events." Activities such as these may release air contaminants in such large amounts that systems designed to meet minimum ventilation standards will not be able to control them.

Unvented combustion appliances are outside the scope of the standard. This includes unvented space heaters and gas cooking appliances. The standard neither prohibits nor permits their installation. You must look to existing codes and manufacturer's instructions for guidance on the use of these appliances.

Section 3 – Definitions

The standard contains definitions for many terms that it uses. See the standard for definitions of:

- acceptable indoor air quality
- air cleaning
- air, exhaust
- air, indoor
- air, outdoor
- air, transfer
- air, ventilation
- air change rate
- balanced system
- bathroom
- climate, hot, humid
- climate, very cold
- conditioned space
- contaminant
- exhaust system
- exhaust flow, net
- habitable space
- heating degree-day
- high-polluting events
- infiltration
- kitchen
- mechanical cooling
- mechanical ventilation
- natural ventilation
- occupiable space
- pressure boundary
- readily accessible
- source
- supply system
- system
- toilet
- utility
- ventilation

Section 4 – Whole-Building Ventilation

Background

As houses have become tighter and we use more man-made materials, we have increased the need to control ventilation to account for indoor moisture and changes in the chemicals in homes. There has always been a need for "fresh air" in all houses and the legal requirement for access to air and light goes back to British Common Law in the eleventh century. In the past, ASHRAE and most building codes have assumed that this "fresh air" would be provided through operable (openable) windows and normal leaks in the building. As houses began to be tightened up over the last twenty-five years due to energy cost concerns and the use of large sheet goods and housewrap, the normal infiltration and exfiltration has been reduced. We can therefore not rely on natural forces to provide ventilation at all times of the year. In the meantime, we have introduced thousands of chemicals into our houses through building materials, finishes, packaging, furniture, carpets, clothing and other products. This is in addition to the allegens, bioaerosols, and occupant generated VOCs that have always been present in the indoor air.

This increase in household pollutants and decrease in unintended infiltration and exfiltration has led to houses that need mechanical or natural ventilation. Natural ventilation using windows and other operable openings can provide adequate ventilation if they are used (which may be more likely when the climate is very temperate). However, there are many reasons why an occupant may choose not to operate the windows, including security, outdoor air quality, or noise. See Section 4.5 below for the exception to the requirement for mechanical ventilation. Most of the new construction in the United States will need mechanical ventilation to comply with Standard 62.2-2010.

4.1 Ventilation Rate

Whole building ventilation is required by the standard in all cases. See the *Background* for this section for information on why this is required. In most cases, ventilation must be provided mechanically by a system of fans (see Section 4.2). Whole-building ventilation is intended to dilute and remove the general air contaminants that are generated by the house, furnishings, wall and floor coverings, and normal occupant activities. The required ventilation rate is based on the size of the house and the number of expected occupants.

The number of occupants is usually unknown at the time of construction, so the standard assumes occupancy based on the number of bedrooms. The standard assumes that the first bedroom will have two occupants, with one additional occupant for each additional bedroom. Studio apartments are assumed to have two occupants.

To find the required ventilation rate in cfm:

$$\text{Cfm} = \text{Floor Area}/100 + (\text{Number of Bedrooms} + 1) \times 7.5$$

The standard requires 7.5 cfm per occupant, plus 0.01 cfm per square foot of floor area (1 cfm per 100 ft^2).

If you are using metric units, the required ventilation in liters/sec is:

L/s = Floor Area /20 + (Number of Bedrooms + 1) x 3.5

where the floor area is in square meters. The required ventilation rates are 0.05 L/s per m^2 (1 L/s per 20 m^2) plus 3.5 L/s per occupant.

Instead of using one of the equations given above, Table 4.1 can be used. There are two versions of the table, one in inch-pound units and the other for metric units. These tables allow the user to find the required ventilation rate directly if they know the floor area and number of bedrooms. The values in the tables are based on the maximum floor area and number of bedrooms from the appropriate row and column.

TABLE 4.1a (I-P) – Ventilation Air Requirements, cfm

Floor Area (ft^2)	**Bedrooms**				
	<2	**2-3**	**4-5**	**6-7**	**>7**
≤1500	30	45	60	75	90
1501-3000	45	60	75	90	105
3001-4500	60	75	90	105	120
4501-6000	75	90	105	120	135
6001-7500	90	105	120	135	150
>7500	105	120	135	150	165

TABLE 4.1b (SI) – Ventilation Air Requirements, L/s

Floor Area (m^2)	**Bedrooms**				
	<2	**2-3**	**4-5**	**6-7**	**>7**
≤139	14	21	28	35	42
139.1-279	21	28	35	42	50
279.1-418	28	35	42	50	57
418.1-557	35	42	50	57	64
557.1-697	42	50	57	64	71
>697	50	57	64	71	78

When selecting a fan to meet this requirement, the fan and its installation must comply with the requirements of Section 7, as discussed later. Also, the required airflow capacity must be rounded up. See Example 4-B.

Exceptions to 4.1: Section 4.1 requires mechanical ventilation in all new dwelling units (houses or apartments), at a rate determined above. The exception to section 4.1, however, provides some conditions when mechanical ventilation is not required:

1. If the building has no air-conditioning and is located in a warm climate, then it is likely that the windows will be open much of the time.

2. If the building is heated or cooled for human occupancy for less than 876 hours per year. (876 hours is 10% of the year, or 36½ days.) An example of this might be a vacation home that is not rented out, and is visited only occasionally. In this case the benefit of reduced exposure to contaminants would be minimal. However, a system still may be considered a benefit by the occupants, due to less irritation due to indoor air contaminants.

If either of these conditions are met, and the local code official (authority having jurisdiction) determines that window operation is a permissible means of providing ventilation, then the installation of a ventilation system is not required.

The locations where Exception 1 applies are defined as climate zone 1 or 2 of the IECC 2009 climate zone map. This covers the gulf coast, Hawaii, southern Arizona, and one county in California, as shown in Figure 8.2 of the standard. It also includes Puerto Rico, the U.S. Pacific Islands and the Virgin Islands. Table 4.2 lists U.S. counties which are located in climate zones 1 and 2.

Example 4-A Determination of Floor Area for Use in Determining Ventilation Rate

Q: A house construction project is planned. How must the floor area be measured to determine the value to be used in finding the required ventilation rate?

A: The standard is silent on how the floor area is measured, although it needs to include all occupiable space, meaning all the spaces which people are expected to regularly enter. You can certainly use the gross floor area found by measuring the outside of the house. You can also measure the inside of all the rooms, including hallways and closets, which will give a lower floor area and therefore a lower ventilation rate requirement. While the standard does not require it, it might be appropriate to use the floor area that will be shown when the house is listed for sale.

Example 4-B Selecting a Fan Capacity

Q: The house I am building has a floor area of 2,240 ft^2 and 3 bedrooms. My calculations come out to 52.4 cfm. Can I use a 50 cfm fan?

A: No, a 50 cfm fan does not meet the standard. You would need to select the next larger fan size, such as a unit rated at 55 cfm or 60 cfm. If you use Table 4.1a to select the fan size, you get 60 cfm.

If the next available fan size provides more than the required airflow, using a multispeed fan may improve the match between provided and required flow. A further option may be to use a larger fan with a timer to cycle the fan, matching the required flow through the use of the calculations in Section 4.5. See Example 4-K.

TABLE 4.2 – U.S. Counties Located in Climate Zones 1 and 2

State, County

Alabama
- Baldwin
- Mobile

Arizona
- La Paz
- Maricopa
- Pima
- Pinal
- Yuma

California
- Imperial

Florida
- All counties

Georgia
- Appling
- Atkinson
- Bacon
- Baker
- Berrien
- Brantley
- Brooks
- Bryan
- Camden
- Charlton
- Chatham
- Clinch
- Colquitt
- Cook
- Decatur
- Echols
- Effingham
- Evans
- Glynn
- Grady
- Jeff Davis
- Lanier
- Liberty
- Long
- Lowndes
- McIntosh
- Miller
- Mitchell
- Pierce
- Seminole
- Tattnall
- Thomas
- Toombs

Georgia (cont.)
- Ware
- Wayne

Louisiana
- Acadia
- Allen
- Ascension
- Assumption
- Avoyelles
- Beauregard
- Calcasieu
- Cameron
- East Baton Rouge
- East Feliciana
- Evangeline
- Iberia
- Iberville
- Jefferson
- Jefferson Davis
- Lafayette
- Lafourche
- Livingston
- Orleans
- Plaquemines
- Pointe Coupee
- Rapides
- St. Bernard
- St. Charles
- St. Helena
- St. James
- St. John the Baptist
- St. Landry
- St. Martin
- St. Mary
- St. Tammany
- Tangipahoa
- Terrebonne
- Vermilion
- Washington
- West Baton Rouge
- West Feliciana

Mississippi
- Hancock
- Harrison
- Jackson
- Pearl River
- Stone

Texas
- Anderson
- Angelina
- Aransas
- Atascosa
- Austin
- Bandera
- Bastrop
- Bee
- Bell
- Bexar
- Bosque
- Brazoria
- Brazos
- Brooks
- Burleson
- Caldwell
- Calhoun
- Cameron
- Chambers
- Cherokee
- Colorado
- Comal
- Coryell
- DeWitt
- Dimmit
- Duval
- Edwards
- Falls
- Fayette
- Fort Bend
- Freestone
- Frio
- Galveston
- Goliad
- Gonzales
- Grimes
- Guadalupe
- Hardin
- Harris
- Hays
- Hidalgo
- Hill
- Houston
- Jackson
- Jasper
- Jefferson
- Jim Hogg

Texas (cont.)
- Jim Wells
- Karnes
- Kenedy
- Kinney
- Kleberg
- La Salle
- Lavaca
- Lee
- Leon
- Liberty
- Limestone
- Live Oak
- Madison
- Matagorda
- Maverick
- McLennan
- McMullen
- Medina
- Milam
- Montgomery
- Newton
- Nueces
- Orange
- Polk
- Real
- Refugio
- Robertson
- San Jacinto
- San Patricio
- Starr
- Travis
- Trinity
- Tyler
- Uvalde
- Val Verde
- Victoria
- Walker
- Waller
- Washington
- Webb
- Wharton
- Willacy
- Williamson
- Wilson
- Zapata
- Zavala

Example 4-C Determining if Mechanical Whole-Building Ventilation is Required – Exception 2

Q1: A vacation home is being built in an area that is a popular vacation destination. The owner plans to use it about 4 weeks per year. To prevent pipe freezing, the heat will be maintained at 45°F whenever the owner is not present. Is mechanical ventilation required?

A1: In this case, the home is occupied for only 672 hours per year. The space is not conditioned *for human occupancy* at other times, so mechanical ventilation is not required.

Q2: What if the owner is considering making the home available for rent during periods when he is not using it? No definite decision has been made, and even if it is made available, it is not known how often it would be rented.

A2: In this case, it might be best to install mechanical ventilation to be on the safe side. However, the determination of the number of hours per year when the house will be in use ("thermally conditioned for human occupancy") only needs to be the best estimate available at the time of construction. You should make your best guess at how much the house will be rented, and if the total (owner occupied plus rented) is less than 10% of the year (about 36½ days), then mechanical ventilation is not required. If the owner will be using it for 672 hours per year, it is likely, though, that if it is rented the total occupancy will exceed 876 hours, so mechanical ventilation should be installed.

4.1.1 Different Occupant Density

The standard assumes that the occupancy of a home will be two occupants for the first bedroom and one additional occupant for each additional bedroom. In some cases, actual occupancy, different from this amount, may be known. In this case, if the actual occupancy is higher than that assumed by the standard, the whole-building ventilation rate must be adjusted accordingly. If the actual occupancy is lower, then the whole-building ventilation rate may be adjusted, but only with the approval of the authority having jurisdiction, normally the local building department. Otherwise the rate based on the default occupancy must be used.

The adjustment in the rate is the difference in the number of occupants, multiplied by the per person ventilation rate of 7.5 cfm (3.5 L/s).

4.1.2 Alternative Ventilation

This section of the standard allows a licensed design professional (an architect or engineer) to approve an alternative method of providing whole-building ventilation. The ventilation rate determined in Section 4.1 must still be provided, but a method other than mechanical fans may be designed.

This section is included in the standard to allow innovative approaches to residential ventilation to be tried. The two criteria – ventilation rate from Section 4.1 and approval by a licensed design professional – are included to provide some assurance that the system will provide adequate ventilation. The types of approaches that may be used are intentionally left open, to encourage new ideas.

Example 4-D Actual Occupancy Different from the Default Value

Q1: I will be building a custom home for a family of 6, which will have a floor area of 2650 ft^2 and 4 bedrooms. How much mechanical whole-building ventilation do I need to provide? The house is located in Maryland.

A1: The default ventilation rate would be 2650 ft^2 x 0.01 cfm/ft^2 + (4 + 1) persons * 7.5 cfm/person = 64 cfm. The default occupancy is 5, 2 for the first bedroom plus 3 for the three additional bedrooms. The actual occupancy will be 6, so the ventilation rate **must** be adjusted by one person, or 7.5 cfm, so the actual required ventilation rate is 71.5 cfm, rounded to 72 cfm.

Q2: A retired couple has asked me to build them a new home in northern Michigan. They will be the only occupants, but the house will be 2390 ft^2 with 3 bedrooms. What whole-building ventilation rate do I need?

A2: The default ventilation rate would be 2390 ft^2 x 0.01 cfm/ft^2 + (3 + 1) persons * 7.5 cfm/person = 53.9 cfm (round to 54 cfm). The default occupancy is 4, 2 for the first bedroom and 2 for the two additional bedrooms. The actual occupancy is 2, which means that you **could** reduce the ventilation rate by 15 cfm, if allowed by the local building inspector or other authority, for a ventilation rate of 38.9 cfm (39 cfm). Note, however, that if system is designed for less than the default occupancy, and in the future the occupancy increases (e.g., the house is sold to a larger family), the ventilation system would no longer meet the requirements of the standard.

Example 4-E Alternative Ventilation Systems

Q1: My brother-in-law, Fred, is a prolific inventor with 7 patents to his name. He has developed a device which he says will give any desired ventilation rate using energy derived from the indoor-outdoor temperature difference. Would use of this device meet the requirements of the standard?

A1: Possibly, but you would need to do a couple of additional things in order to meet the standard. First, you don't say whether Fred has a professional license, but the device and its installation in the house must be approved by a Professional Engineer or licensed Architect. Second, the idea that the device is driven by indoor-outdoor temperature differences is possibly a problem because when the outdoor temperature is the same as the indoor temperature, the device may not provide any ventilation. Such a configuration would not comply, because an intermittently operated ventilation system must be controlled by a timer. If Fred's device can provide a constant ventilation rate, regardless of the outdoor temperature, and he can convince a registered engineer or architect to approve it, then the system would meet the standard.

Q2: I am a registered architect designing a custom home. I would like to use a thermal stack device to provide natural ventilation, but I know that when the outdoor air temperature is close to or warmer than the indoor temperature, the thermal stack may not provide sufficient ventilation. I have done some modeling studies, and am convinced that the stack will provide adequate ventilation anytime the outdoor temperature is below 55°F. Can I use the stack in combination with a mechanical ventilation system that operates only when the temperature is above 55°F?

A2: Yes, the system you describe would comply with the standard. As long as the system with both the thermal stack and the fan provides the ventilation rate determined in Section 4.1 at all outdoor temperatures, and you, as a "licensed design professional," have approved the design, then it complies with the standard.

4.1.3 Infiltration Credit

The purpose of 4.1.3 is to allow a reduction in the required ventilation rates in section 4.1. The rates given in section 4.1 are the basis for selecting ventilation system size. Table 4.1a assumes, in addition to the mechanical ventilation, an infiltration rate of 2 cfm per 100 square feet of occupiable floor space. (Table 4.1b assumes infiltration of 10 L/s per 100 m^2.) This value is based on a fairly tight house, built with good attention to air sealing.

If it is determined by testing that the house has a higher infiltration rate, then the required mechanical ventilation rate can be reduced by one-half the difference between the measured infiltration and the default infiltration.

The actual infiltration must be determined using ASHRAE Standard 136[4]. The airflow rate of the mechanical ventilation system is then:

Cfm = Cfm from Table – ½ x (Measured Infiltration in cfm/ft^2 – 0.02 cfm/ft^2) x Floor Area in ft^2, or

L/s = L/s from Table - ½ x (Measured Infiltration in L/s/m^2 – 0.10 L/s/m^2) x Floor Area in m^2

The reduction in the mechanical ventilation rate is only one-half the infiltration rate difference because infiltration rate varies based on the indoor-outdoor temperature difference, the wind velocity and the size and location of air leaks in the building enclosure.

Example 4-F Infiltration Credit

Q1: I am renovating a three bedroom, 2,400 ft^2 house (Figure 4-3) which has a measured air leakage of 72 cfm. Can I reduce the size of the mechanical ventilation system? If so, what size system do I need?

A1: From Table 4.1a, the mechanical ventilation system is required to provide 60 cfm. The standard assumes an infiltration rate of:

2cfm x floor area/100

2 x 2400/100 = 48cfm of infiltration

Therefore the total ventilation rate (mechanical plus infiltration) is 108 cfm (60 cfm mechanical + 48 cfm of assumed infiltration).

The measured infiltration rate, according to ASHRAE Standard 136, is 72 cfm. This is 3 cfm/100 ft^2 (72 cfm/ 2400 ft^2 x 100). Since this is larger than the default infiltration rate of 2 cfm/100 ft^2, the mechanical ventilation rate can be reduced. The adjusted mechanical ventilation rate is:

Cfm = 60 Cfm (from Table) – ½ x (3 cfm/100 ft^2 – 2 cfm/100 ft^2) x 2400 ft^2/100

Cfm = 48 cfm of mechanical ventilation

This can also be found using the two infiltration rates:

Cfm = 60 Cfm (from Table) - ½ x (72 cfm – 48 cfm)

Cfm = 48 cfm

In this case, because the infiltration rate has been measured to be 24 cfm more than the default infiltration credit, the mechanical ventilation rate can be reduced by half that amount, 12 cfm, so the mechanical ventilation system only needs to provide 48 cfm instead of 60 cfm.

Q2: For the 2,400 ft^2 house mentioned above, what would the measured leakage need to be in order for mechanical ventilation to not be required at all?

A2: As described above, the standard assumes that 108 cfm of ventilation is being provided by a mixture of mechanical ventilation and infiltration. Since excess measured leakage is credited at 50% toward offsetting installed mechanical ventilation systems, 168 cfm of measured leakage must be present to completely offset the requirement for mechanical ventilation.

Q3: I am renovating a two bedroom, 1,000 ft^2 apartment (Figure 4-4). I have used ASHRAE 136 to find a measured air leakage of 30 cfm (3 cfm/100 ft^2). Should I be concerned that the Scope of Standard 136 says that the rates calculated by the standard "apply only to detached single-family dwellings"?

A3: In terms of compliance with Standard 62.2, no, you do not need to be concerned. Using a rate as determined by Standard 136 is required by Standard 62.2, regardless of the scope limitations in Standard 136.

Q4: For the two bedroom, 1,000 ft^2 apartment, what size mechanical ventilation system do I need?

A4: The required mechanical ventilation rate from Table 4.1a is 45 cfm. The required ventilation when the infiltration credit is used is then:

$$\text{Cfm} = 45 \text{ Cfm (from Table)} - \tfrac{1}{2} \times (3 \text{ cfm}/100 \text{ ft}^2 - 2 \text{ cfm}/100 \text{ ft}^2) \times 1000 \text{ ft}^2/100$$

$$\text{Cfm} = 40 \text{ cfm of mechanical ventilation}$$

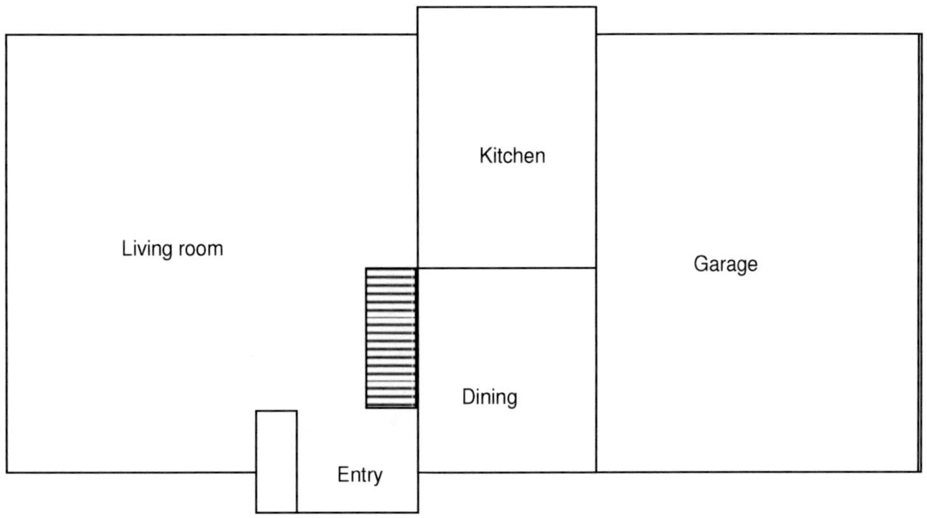

First Floor

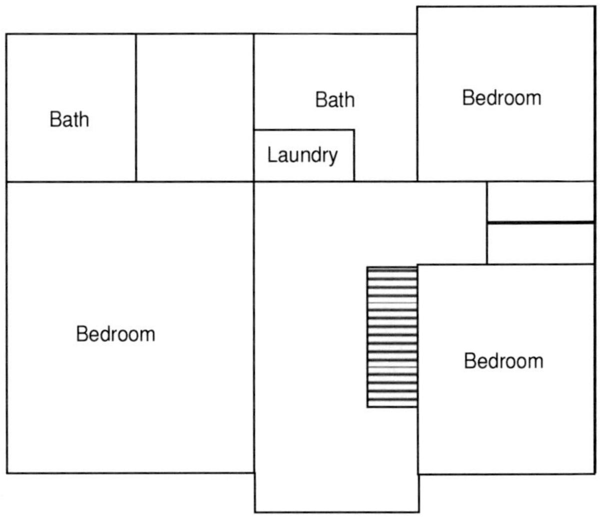

Second Floor

Figure 4.3 – Floor plan for example 2400 square foot house

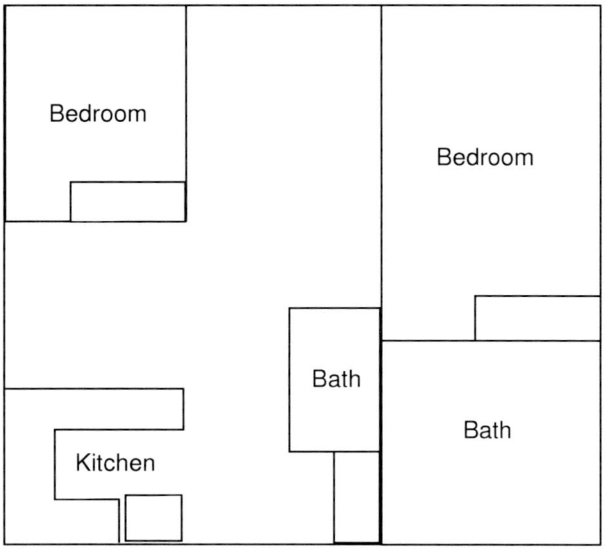

Figure 4.4 – Floor plan for example 1000 square foot apartment

4.2 System Type

Section 4.2 requires that the home include a whole-building ventilation system designed to meet the ventilation rates found in Section 4.1. The system includes:

- the fan or fans,
- the ducts and fittings,
- the terminal device, and
- the controls for the system.

This ventilation system is intended to provide adequate mechanical ventilation whenever the home is **occupiable,** not just when occupied. This means the system must be capable of providing the desired ventilation 24 hours a day, 365 days a year. This can be met with continuous or intermittent ventilation. If by intermittent, then increased fan airflow must compensate for the reduced run time. Section 4.4 below will address how to calculate the increase in flow rate that would be required to decrease the operating time.

The standard allows the designer to choose a system that is appropriate for the climate and building type. It makes no judgment on what is best where. ASHRAE Guideline 24-2008 does provide some guidance in Chapter 10. The ventilation system can use:

- supply ventilation only,
- exhaust ventilation only, or
- a combination of supply and exhaust.

A combination system may or may not include heat and/or energy recovery. Generally, the designer should avoid excessive pressurization (excessive supply ventilation) in severe cold climates and excessive depressurization (excessive exhaust ventilation) in hot, humid climates. The standard sets limits on this in Section 4.5 below. In addition to the whole-building ventilation system, Section 5 of the standard requires local exhaust systems.

Possible options for a 2,800 square foot, three bedroom house needing 60 cfm of outdoor air would include

- a 60 cfm quiet bath exhaust fan
- two 30 cfm bath fans
- a single inline fan set to 60 cfm
- an inline fan with two 30 cfm pickups
- an inline fan or blending ventilator used to bring in 60 cfm (supply)
- an HRV or ERV with a continuous setting of 60 cfm
- a duct connection to the outside from the return air plenum sized to provide 60 cfm while the air handler operates continuously
- a duct connection sized to bring in 120 cfm with the air handler operating at least half the time.

Supply ventilation can be provided through dedicated supply fans or through the central HVAC system air handler. Dedicated supply fans can be mounted on the exterior of the house or dwelling unit or can be mounted inline in the garage, attic, basement, or mechanical room to pull in outdoor air and supply it to the dwelling unit either directly or through the HVAC ducting. Some supply fans include a filter and/or a return from the house to provide some air cleaning or tempering. There are a variety of products that can clean the air or dehumidify it before it is introduced into the house.

The HVAC air handler can be used as a supply ventilation system with the addition of a direct duct connection to the return air duct from the outside. See Figure 4.5 for an example of this type of system. This strategy uses the negative pressure in the return plenum to pull in outdoor air and then push it through the air handler. Design considerations include:

- Furnace manufacturers have minimum return air temperature requirements (typically 55°F). Usually, the mixed air temperature would be higher than this, but it is theoretically possible to get a lower mixed air temperature in very cold climates with high outside air flow rates.
- If the "fan-on" setting of a conventional thermostat is used, excessive fan energy will be used. Use of a timer control could lessen this problem, if the system meets the ventilation airflow as calculated under section 4.5, below. For example, if the timer operates the fan for 20 minutes per hour, then the delivered ventilation rate when the fan is operating must be three times greater than the calculated continuous ventilation rate from Section 4.1.
- Proper design and implementation of this system is important. Proper outside airflow should be verified, and timer controls should be used and set to operate the system the designed minimum amount of time.

For either type of supply system, the location of the inlet must comply with Section 6.8 which prohibits their placement within specified distances from specific contaminant sources. In addition to the sources listed in Section 6.8, placement of the outdoor air inlet should also avoid sources of contaminants or odors, such as the garage, barbeque areas, and chimneys, although this is not a requirement of the standard.

Care must also be taken to avoid uncomfortable conditions that might be created by introducing too much outdoor air into one location.

Supply ventilation systems depend on normal leaks in the building envelope and exhaust fans as "escape routes" for the indoor air.

Exhaust ventilation can be provided by quiet ceiling-mounted bath fans or remote-mounted inline or exterior-mounted fans. When selecting an exhaust fan, the sone rating for a local exhaust fan must be 3 sones or less, unless the local exhaust fan is located remotely (no sone requirement), or is also used for providing continuous or whole-building ventilation (must be 1.0 sone or less). Many high quality bath fans are available in the 30 to 150 cfm range that are rated at 1.0 sone or less (See Section 7.1 below). Most dwelling units will only need 50 to 90 cfm of ventilation, so it is possible to use a fan rated at 0.5 sone or less. This can be a dedicated whole-building ventilation fan or it can be one of the bath fans in the house that is used for both whole-building ventilation and local exhaust. See Figure 4.6 for an example system. The standard explicitly allows the use of a local exhaust fan as part of the ventilation system.

Inline fans, either single pickup or multipoint pickup, can be a very effective method of providing quiet exhaust ventilation from one or several bathrooms. As discussed above, inline fans can be mounted remotely in the garage, attic, basement, or mechanical room. Exterior-mounted fans can be mounted on the exterior wall or on the roof. See Figure 4.7 for an example of this type of system. There is no sound rating required for remote fans as long as there is at least four feet of duct between the pickup grille and the fan. Exhaust ventilation systems depend on the normal leaks in the exterior of the home for the introduction of the outdoor air. Because the outdoor makeup air comes from small leakage paths, this outdoor air is generally not noticed as a draft by the occupant. Design considerations include:

- Because the source of incoming outdoor air is not known, particular attention must be given to avoiding drawing air from or through areas with contaminated air such as garages, crawlspaces, attics, and slabs or walls with soil contact. In multi-family dwellings, special attention must be given to air sealing between adjacent units.
- Exhaust ventilation contributes to depressurization of the building with respect to outdoors. Attention should be given to depressurization and potential back-drafting of combustion systems.
- Using an isolated bathroom exhaust fan, such as a fan located in a master bathroom, to provide whole-building ventilation may not achieve distribution of ventilation air.

Combination ventilation systems can be used as well to meet the ventilation requirement. Often, combination systems are integrated devices that have two airstreams to provide both exhaust and supply ventilation. The two airstreams are normally fairly equal, so making the system "balanced" to avoid either pressurization or depressurization, but the standard does not require this. Some integrated balanced systems can provide heat or energy recovery from one air stream into the other by using a Heat Recovery core or wheel to move heat or an Energy Recovery core or wheel to recover both sensible heat and the latent heat of evaporation or condensation. Heat and Energy Recovery Ventilators (HRVs and ERVs) are used in cold climates to preheat the incoming air or in hot climates to cool the incoming air with the cooled air from the house. Energy Recovery Ventilators (ERVs) are able to recover the sensible heat from the temperature difference as well as the latent heat of evaporation when the moisture condenses from the warmer air stream as it is cooled by the other air stream. ERV's also transfer moisture between the air streams, from higher water vapor concentration to lower. See Figure 4.8 for an example system configuration with an HRV or ERV connected to the central air handler for whole-building ventilation, continuously operating exhaust pickups in the bathrooms, and a range hood in the kitchen.

Design considerations include:

- Designs that remove air from the HVAC duct system and supply outdoor air to that same duct system require that the HVAC system fan operate whenever the balanced system operates to avoid short-circuiting. This can significantly increase the energy impact of this ventilation strategy.

- Designs that de-couple one or both sides of the balanced system's ducting from the home's HVAC system ductwork do not need to rely on the HVAC fan.

Combination systems can also consist of a mixture of supply fans and exhaust fans. It may be as simple as a quiet continuous bathroom exhaust fan matched to an outdoor air connection that introduces air into the return air plenum of a continuously-operating air handler. Dedicated systems can also be used that include filtration or tempering of the incoming air. In any case, the two airstreams are often the same nominal cfm to maintain a balanced flow of supply and exhaust air.

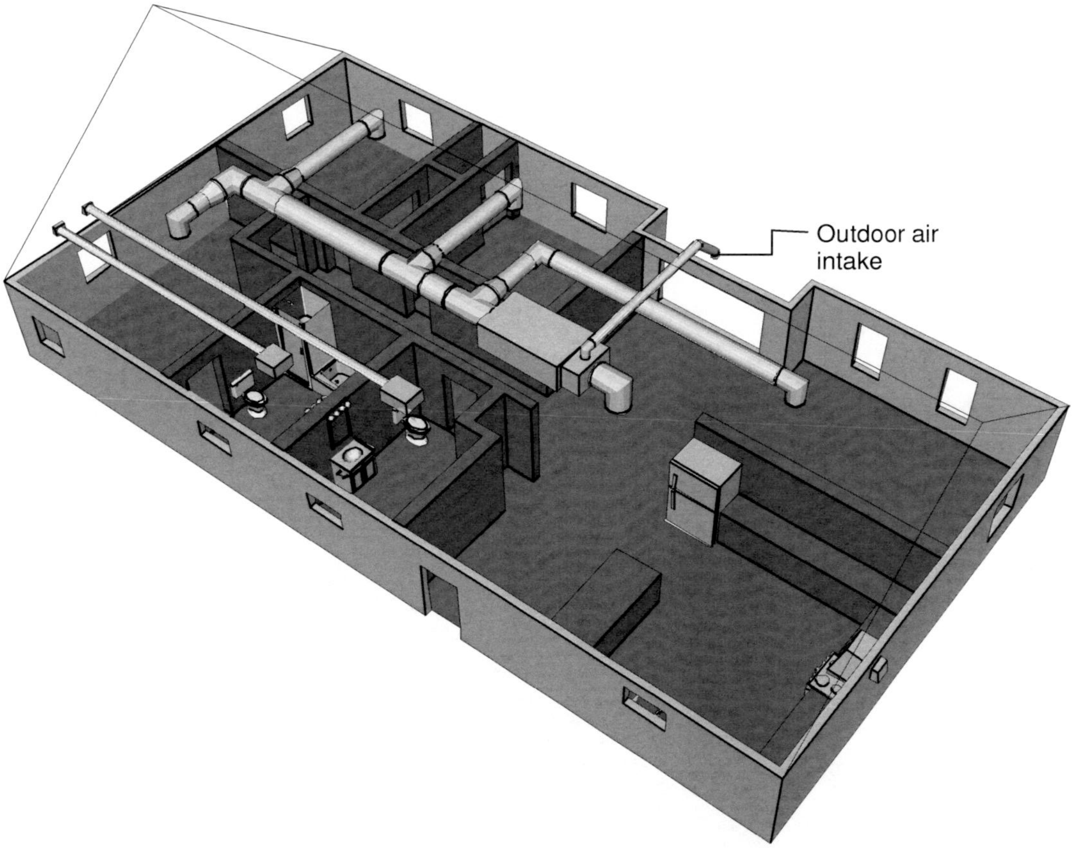

Figure 4.5 Example Supply Only Ventilation System using an Outdoor Air Duct to the Return Side of the Central Air Handling Unit

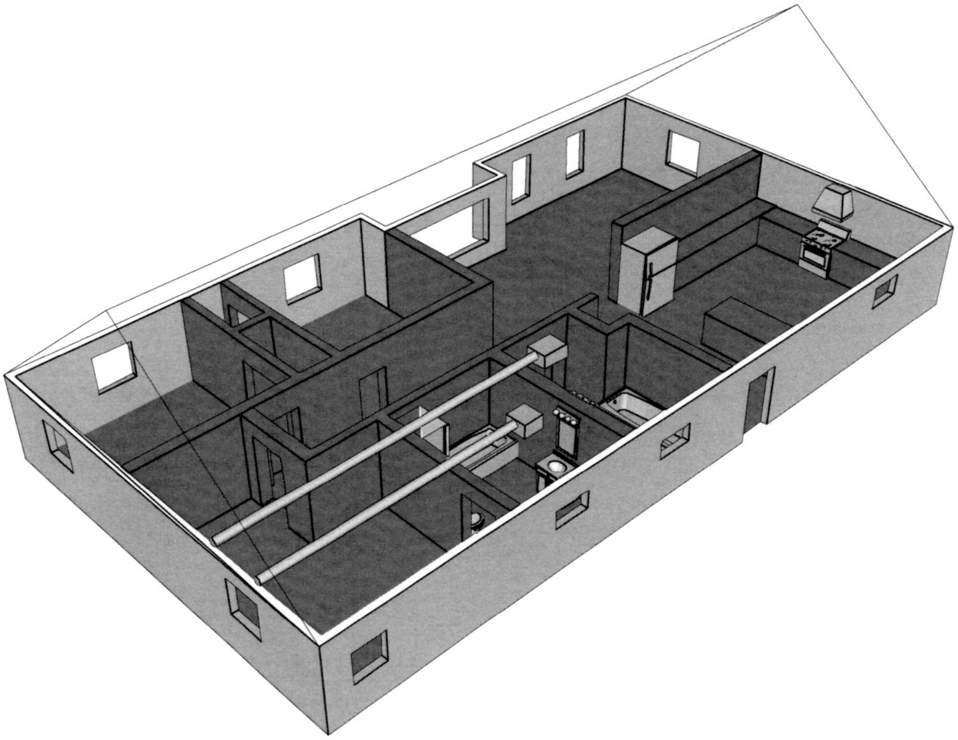

Figure 4.6 – Example Exhaust Only Ventilation System using Dedicated Exhaust Fans

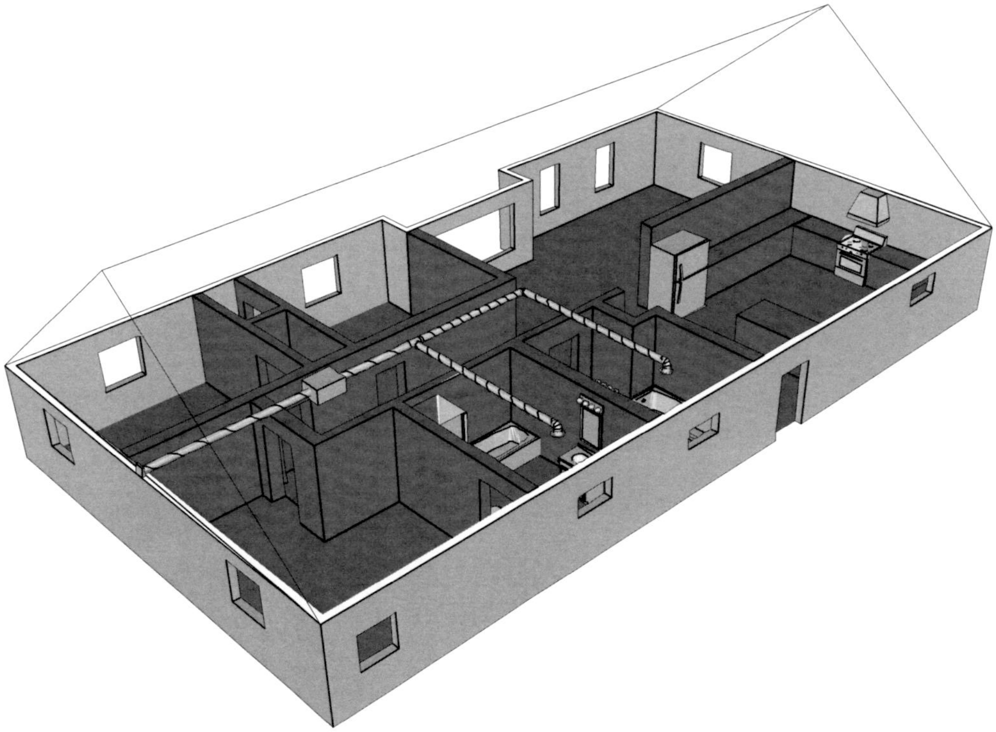

Figure 4.7 – Example Exhaust Only Ventilation System using a Central Inline Exhaust Fan with Multiple Pickups

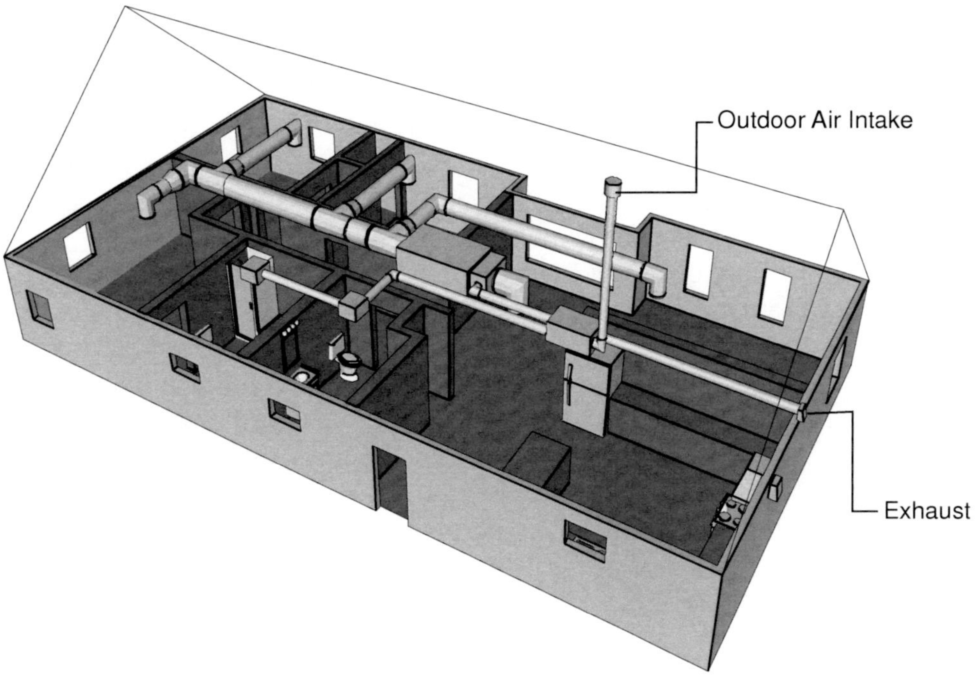

Figure 4.8 Example Ventilation System Using an HRV/ERV Attached to the Central Air Handler with Local Exhaust.

Figure 4.9 – Flow Hood Being Used to Measure Airflow from a Supply Outlet

4.3 Airflow Measurement

The 2010 edition of ASHRAE 62.2 adds a requirement that the airflow of the whole-building ventilation system has to be measured after installation to verify compliance with the standard. The airflow can be measured with a flow hood, a flow grid, an airspeed meter, or other device that is allowed by the local code official (authority having jurisdiction). This applies to all types of whole-building ventilation whether continuous systems or intermittent systems using larger fans and airflows. It does not necessarily apply to local exhaust fans required under Section 5 - Section 5 includes requirements similar to this, but they can be avoided through adherence to prescriptive duct sizing.

If the equipment or system being measured has multiple operating modes, an airflow measurement must be made with the system operating in any mode that is intended to be used to meet Standard 62.2.

4.4 Control and Operation

Section 4.4 has two main requirements, one of which has an exception. First, the standard requires that the ventilation system have an override control which is readily accessible to the occupants. The "fan-on" switch on a typical thermostat controlling the HVAC system and the wall switch for an exhaust fan are both explicitly allowed as acceptable controls. This control must be "readily accessible" as defined in Section 3, which is similar to the definition in most building codes. This means it must be capable of being accessed quickly and easily without having to remove panels or doors. It can be as simple as a labeled wall switch by the electrical panel. It may be integrated in a labeled wall-mounted control or in the air moving device that requires the removal of the cover plate, but it cannot be buried in the insulation in the attic or the inside of the fan. The occupant must be able to modify the settings or override the system if he or she so chooses. While the standard requires the whole-building ventilation system, the occupant has the option of choosing not to operate the system.

The second requirement in the section has to do with the controls of the ventilation system. Section 4.5 is written for ventilation systems that will operate continuously, but it also includes an exception which allows the use of intermittently operating fans. If intermittent fans are used, they must be controlled by a timer, and they must have an increased airflow rate to compensate for the off time. The system must operate for at least 2.4 hours per day (10% of the time), and must operate at least once per day. The airflow rate of the intermittent system is determined in Section 4.5. If it is intended to operate intermittently, all components of the system must be designed for the increased airflow rate.

Time-of-day timers or duty cycle timers can be used to provide intermittent whole-building ventilation. Manual crank timers cannot be used, since the system must have the capacity to operate automatically without intervention by the occupant. Some controls "look back" over a set time interval to see if the air handler has already operated for heating or cooling before it turns on the air handler for ventilation only operation.

Example 4-G Control Options

Q: I plan to use a bathroom exhaust fan to provide the ventilation for a house. The fan is designed to be operated by a typical wall switch. Do I need to put a label on the wall plate to comply with the requirement that controls be "appropriately labeled"?

A: Yes. If the exhaust fan were serving only the local exhaust requirement for the bathroom, then a label would not be required. Since the fan is providing whole-building ventilation, a label is needed to inform the occupant of that.

4.5 Delivered Ventilation

The ventilation rate determined in Section 4.1 is the rate required from a continuously operating system. Section 4.5 clarifies the ventilation rate provided by a combination ventilation system, and also specifies the airflow required from an intermittently operating ventilation system.

The first part of Section 4.5 simply states that when a combination ventilation system is used, meaning that both supply and exhaust fans are installed, the ventilation rate is the larger of the total supply airflow or the total exhaust airflow. The airflow rates of the supply and exhaust fans cannot be added together to get credit for a higher ventilation rate.

Example 4-H Thermostatic Control

Q: I plan to provide ventilation air by a duct from the return side of the central air handler to the outdoors. Ventilation will be provided whenever the air handler operates. According to my estimates, the system will run on calls for heating and cooling about 40% of the time, averaged over the year. If I provide a safety factor and assume that it only runs 25% of the time, and size the airflow accordingly, can I allow the system to run under thermostatic control?

A: No. A system under thermostatic control will go through periods with little or no operation when the outdoor temperature is near the indoor setpoint, or if the system is in setback mode. An intermittently operating ventilation system MUST be controlled by a timer in order to assure that adequate ventilation is provided regardless of outdoor conditions.

As mentioned in the text, there are timer based controls available which keep track of when the system operates thermostatically, and only turn on the fan when it has not already operated enough to provide the desired ventilation.

Certain system designs that rely on the heating and cooling system to provide ventilation have a significant energy penalty. This would occur if the design of the system is such that the system fan must run continuously or if the system does not have a damper that prevents over-ventilation during extreme weather conditions. Controllers exist that can monitor and cycle the system fan and, if properly used, can significantly reduce or eliminate these forms of energy penalty.

The second part of Section 4.5 states that the airflow values from Section 4.1 shall be provided during each hour of operation, implying that a continuously operating ventilation system is installed. There is an exception to this, however. If the ventilation system runs intermittently based on a timer (see Section 4.4 on Control and Operation), then an equation provides a means of calculating a larger airflow that would be provided during the "on-cycle."

Exception to 4.5: There is an exception to 4.5. Ventilation fans can be run intermittently rather than continuously. To qualify for this exception the fan powered airflow must be larger than that required by Table 4.1a (IP). Greater airflow is needed to account for the time the fan is turned off and the fact that contaminant concentrations will be higher during the time the fan is turned off. The exception provides an equation to calculate the size of the fan flow required:

$$Q_f = Q_r/(\varepsilon f)$$

Where

Q_f = fan flow rate

Q_r = ventilation air requirement (Table 4.1)

ε = ventilation effectiveness (Table 4.2)

f = fractional on-time.

The equation is designed to provide fan flows that will result in contaminant exposures that are equivalent to those provided by a continuously operating ventilation system that meets the flows given in Table 4.1. It results in significantly larger fan flow requirements when the cycle time is more than 8 hours, especially when the fractional on-time is low.

In addition, the standard requires that the fractional on-time must be at least 0.1, and the Cycle Time cannot be more than 24 hours. In other words, the ventilation system must run at least once per day.

Example 4-I Delivered Ventilation

Q: In the 2,400 square foot example house, the exhaust fans running continuously in the two bathrooms provide a total exhaust flow rate of 40 cfm in accordance with Table 5.2. Make-up air is drawn into the building through openings in the enclosure. However, 60 cfm is required by Table 4.1a (IP). What are the options for providing the required 60 cfm?

A: The required 60 cfm could be provided either by increasing the exhaust flow by 20 cfm or by adding a ventilation system that blows 60 cfm of outdoor air into the building. It cannot be achieved by using a make-up air fan blowing 20 cfm into the house.

TABLE 4.2 – Ventilation Effectiveness for Intermittent Fans

Fractional On-Time, f	Cycle Time, T_{cyc} (hr) 0-4	4-8	8-12	12-24
0.1	1.00	0.79	Not Allowed	Not Allowed
0.2	1.00	0.84	0.56	Not Allowed
0.3	1.00	0.89	0.71	Not Allowed
0.4	1.00	0.92	0.81	0.20
0.5	1.00	0.94	0.87	0.52
0.6	1.00	0.97	0.92	0.73
0.7	1.00	0.98	0.96	0.86
0.8	1.00	0.99	0.98	0.94
0.9	1.00	1.00	1.00	0.99
1.0	1.00	1.00	1.00	1.0

Example 4-J Flowrate for Intermittent Fan

Q1: For the example 2,400 ft^2 house the required ventilation rate is 60 cfm. If the ventilation fan runs 80% of the time, with a cycle time of 1 hour, what must the airflow rate be?

A1: Since f is 0.8 (80%) and the cycle time is less than 4 hours, then the ventilation effectiveness, ε, is 1. Q_f equals 60/(0.8 x 1) = 75 cfm. This is a fairly small increase in fan size.

Q2: For the same house, if the fan runs half the day (12 hours per day), what is the required airflow?

A2: The fractional on-time is 0.5 (50%) and the cycle time is 24 hours, so ε is 0.52 from Table 4.2. The fan size, Q_f = 60/(0.52 x 0.5) = 231 cfm. This is a much larger increase in fan size.

Q3: For the 1,000 ft^2 apartment example, the flow required by Section 4.1 is 45 cfm. If the ventilation fan runs 20 minutes on and 10 minutes off, what is the required fan size?

A3: Fractional on-time is 0.67 (67%). [f = on-time/total time = 20/(20 + 10)] Since the cycle time is 30 minutes, ε is 1.0. The fan size, Q_f = 45/(0.67 x 1.0) = 67.1 cfm, which rounds to 68 cfm.

Q4: For the same apartment, if the fan runs 8 hours on and 4 hours off, what flow rate is required?

A4: Fractional on-time is again 0.67 (67%), but the cycle time is now 12 hours. Table 4.2 provides ε as 0.73. Q_f = 45/(0.67 x 0.73) = 92 cfm.

Q5: For the 2,400 ft^2 house, I have an electronic timer system. I would like to have the system run only 2 hours in the morning and 8 hours in the evening (6 a.m. – 8 a.m. and 4 p.m. to midnight). I can set the timer to operate the fan for 1 minute every hour. What flow rate do I need?

A5: Forget about the 1 minute every hour. The system must have a fractional on-time of at least 0.1, which is at least 6 minutes per hour, if it ran every hour. However, in this case, the cycle time is 24 hours, and the minimum allowed fractional on-time from Table 4.2 is 0.4. This 0.4 is applied to the total day, so the required on time is 0.4 x 60 minutes per hour x 24 hours per day = 576 minutes. When this time is distributed over the desired 10 hours of operation, the result is 57 minutes and 36 seconds of operation each of those hours.

For a cycle time of 24 hours, with a fractional on-time of 0.4, the effectiveness is 0.20. The required flow rate, Q_f, is 60/(0.4 x 0.2) = 750 cfm. This is a very large flow rate, and indicates that the proposed operating strategy can be substantially improved.

4.6 Restrictions on System Type

Background

When humid air comes into contact with surfaces which are colder than the dewpoint of the air, water will condense onto the cold surface. If this process continues such that the surface is wet for extended periods, mold and mildew will grow on the surface. Mold and mildew on surfaces can result in indoor air quality problems.

When very humid outdoor air is drawn into an air-conditioned building through the envelope, if the outdoor dewpoint temperature is higher than the indoor space temperature, then at some point it will come into contact with building components that are below the dewpoint. Condensation may then occur. Often, this happens inside the building's walls, where mold growth could go undetected for extended periods, until the result is structural damage or health effects on the occupants.

While not as common, a similar effect can occur during the heating season in cold climates, particularly when a humidifier is used. If the warm, humid indoor air is forced out through the envelope, at some point it reaches a cold building component where condensation may occur.

Section 4.6 places some restrictions on whole-building net exhaust for hot, humid climates and net mechanical supply for severe cold climates. The purpose of the restriction is to reduce the chance of condensation in the enclosure (e.g. walls, floors, ceilings, attics). Table 8.2 in the standard lists U.S. cities which are located in hot, humid climates, and Table 8.3 in the standard lists very cold U.S. cities. These are not exhaustive lists, of course, but if the building site is near or between cities listed, then the appropriate restriction (hot, humid or very cold) must be applied. Figure 8.1 in the standard shows a map of North America where the hot, humid and very cold climates are shown. Hawaii (not shown on the map) is considered a hot, humid climate. Note, however, that some parts of Hawaii are quite dry or cold (the western part of the Big Island is very dry).

Example 4-K Run Time for Intermittent Fan

Q: In Example 4-B, the required ventilation rate for a residence was calculated to be 52.4 cfm. A 50 cfm fan does not meet the standard, and the next available fan size is 65 cfm. If a timer system is installed to cycle the fan, how much time each hour must it operate to meet the standard?

A: Since it is assumed that the fan will cycle once per hour, the ventilation effectiveness, ε, is 1. The fractional on time is $Q_r/(\varepsilon\ Q_f)$ which is 52.4/(1 x 65) = 0.806 hours. Multiplying by 60 gives an on time of 48.4 minutes per hour, or 48 minutes and 24 seconds.

Unless the house is extremely airtight, a ventilation system that is sized according to Section 4.1 will not create excessive pressurization or depressurization. If the ventilation system is oversized or if there are fans in the house in addition to the ventilation system, then excessive pressurization or depressurization can occur. A typical example is when a commercial grade kitchen exhaust fan is installed without providing sufficient makeup air.

4.6.1 Hot, Humid Climates. (Note: This section has been removed from future versions of the standard via Addendum 62.2g.)

In hot humid climates the air outdoors is usually warmer and more humid than the air indoors. To reduce problems caused by drawing hot, humid air into a building, the amount of whole-building net exhaust air is limited to 7.5 cfm per 100 square feet. If the system is a combination with both supply and exhaust air streams, this restriction applies only if the exhaust airflow exceeds the supply airflow by 7.5 cfm per 100 ft^2.

4.6.2 Very Cold Climates. (Note: This section has been removed from future versions of the standard via Addendum 62.2g.)

In very cold climates the indoor air is usually warm and can be humid compared to the outdoor air. In this case the restriction applies to mechanical supply ventilation systems that would pressurize the house, pushing warm air into the enclosure. The net mechanical supply flows cannot be greater than 7.5 cfm per 100 square feet of floor area.

Exception to 4.6.1 and 4.6.2: Within the hot humid or severe cold climates an exception to the restrictions can be made if the authority having jurisdiction approves the design as being moisture resistant. General guidance for making a design moisture resistant includes:

- Make the enclosure very airtight.
- Use porous, mold-inhibiting materials in the assemblies (it's harder to get condensation in porous materials than on impermeable ones).
- Use insulated sheathing in very cold climates.

Example 4-L Limit on Exhaust Only Ventilation System

Q1: For the 2,400 ft^2, three bedroom house used previously in Example 4-F (Figure 4-3), Section 4.1 requires 60 cfm. If the house is located in Miami, can I use an exhaust only ventilation system?

A1: The restriction only applies when the airflow exceeds 7.5 cfm per 100 ft^2. For the example house, this is 7.5 x 2400/100 = 180 cfm of net exhaust air. This is larger than 60 cfm, so the restriction does not apply.

Q2: For the same house, I plan to use an intermittent ventilation system that will operate 50% of the time. From Section 4.1 the required ventilation airflow is 60 cfm. From Section 4.5, the required fan airflow is 231 cfm (see Example 4-J, Q2). This airflow exceeds the 180 cfm limit. What can I do so the system will comply with the standard?

A2: There are several options. The on-time of the intermittent system can be increased to 60% or greater. If it were 60%, the required fan flow would be 137 cfm (60 cfm/0.6/0.73), which would not trigger the restriction. Alternatively, a supply only system or a combination system providing at least 51 cfm of supply air can be used so that the net exhaust flow is less than 180 cfm.

Example 4-M Limit on Supply Only Ventilation System

Q: For the 1,000 ft2, two bedroom apartment used in Example 4-F (Figure 4-4), Section 4.1 requires 45 cfm of ventilation air. If the apartment is located in Minneapolis, can I use a supply only ventilation system?

A: Yes, for two reasons. First, Minneapolis is not included in the very cold climate, so supply only ventilation can be used at any airflow rate. Table 8.3 lists very cold U.S. Cities.

Second, even if the apartment were located in International Falls or Duluth, both very cold cities, the restriction only applies when the airflow exceeds 7.5 cfm per 100 ft^2. For the example apartment, this is 7.5 x 1000/100 = 75 cfm of net supply air. The restriction does not apply.

Section 5 Local Exhaust

5.1 Local Mechanical Exhaust

Background

Local exhaust (sometimes called spot ventilation) has long been required for bathrooms and kitchens to deal with moisture and odors at the source. Building codes have required an openable window or an exhaust fan in baths for many years and have generally required kitchen exhaust either directly through a fan or indirectly through a ventless range hood and an openable window. The standard recognizes the limitations of these indirect methods of providing ventilation to reduce moisture and odors and requires that these spaces be mechanically ventilated to the outside even if windows are present. As we have built tighter homes with more insulation, the relative humidity in the home has increased and the potential for condensation on cool or cold surfaces has increased as well. The presence of moisture condensation has been a leading cause of mold and mildew in both new and existing construction. The occurrence of asthma has also increased as the interior relative humidity has gotten higher. Therefore, it has become more important to remove the moisture from bathing and cooking right at the source.

The standard requires that each kitchen and bathroom have a local exhaust system installed. Usually this will mean an exhaust fan, although systems which exhaust from multiple rooms are allowed. The standard defines kitchens as any room containing cooking appliances, and bathrooms are rooms containing a bathtub, shower, spa, or other similar source of moisture. Note that a room containing only a toilet and sink is not required by the standard to have mechanical exhaust.

The standard allows the designer the choice of intermittent operation or continuous operation of the local exhaust ventilation system. Intermittent systems must meet requirements in Section 5.2 of the standard, and continuous systems must meet the requirements in Section 5.3. As will be seen in Sections 5.2.2 and 5.3.2 below, the ventilation rates are different because the ventilation effectiveness is different for an intermittent fan than a continuous fan. Just like whole-building ventilation in Section 4.5, a larger fan is required for intermittent use rather than a smaller fan that is operated continuously. An exception to Section 5.1 allows a licensed design professional to design or approve some other methodology of providing the required level of local ventilation without using a fan.

Example 5-A Local Exhaust Required for Half Bath

Q1: I am building a house with 2½ baths. The half bath consists of a room with a toilet and sink. Is local exhaust required for the half bath?

A1: .No. Local exhaust is required only for bathrooms, which are defined by the standard as rooms with a bathtub, shower, spa or some other similar source of moisture. This does not include a simple sink for occasional hand washing.

Q2: The master bath suite in a house has a bathroom with a shower, spa and sinks. The toilet is in a separate, adjacent room with a full door. Where do I need to install local exhaust fans? .

A2: The standard only requires local exhaust in the bathroom, not the separate toilet room.

Building codes may require that fans used for kitchen range hood ventilation must be safety-rated by UL or some other testing agency for that location and/or application. Typically, these requirements address the fire safety issues of fans placed within an area defined by a set of lines at 45° outward and upward from the cook top. Few "bath" fans will have this rating and cannot be used in this area of the kitchen ceiling.

Example 5-B Acceptable Ceiling Fan for Cooking Area

Q: I purchased a ceiling fan, planning to use it to provide intermittent exhaust in a kitchen. I don't see a label indicating that it is approved for use in a cooking area. How do I know if it is acceptable?

A: Check the packaging for a label or printing that says, **"Not for use in cooking areas,"** or something similar. There is no labeling that says a fan is approved for use in a cooking area, only if it is not approved.

5.2 Intermittent Local Exhaust

The standard requires that intermittent exhaust fans be designed to be operated by the occupant. This usually means that a wall switch or some other type of control is accessible and obvious. There is no requirement stating where the control or switch must to be located. Obviously, bath fan controls are generally located next to the light switch and range hood or downdraft fan controls are generally integrated into the hood or mounted on the wall or counter. Bathrooms can use a variety of exhaust strategies. They can have typical ceiling bath fans or just one or two pickups for remote inline or exterior-mounted fans or HRV/ERV products. They can be integrated with the whole-building ventilation system to provide both functions. Kitchens can have range hoods, downdraft exhausts, ceiling exhaust fans, wall fans, or pickups for remote inline or exterior-mounted fans. Generally, HRV/ERV manufacturers will not allow kitchen pickups to avoid the issue of grease buildup in the heat exchange core. Building codes typically require that the kitchen exhaust must be exhausted through metal duct for fire safety.

Example 5-C Ducting Kitchen Exhaust to the Outdoors

Q1: How do I know what kind of duct I need to use. I've been using recirculating hoods my entire career, now I need to vent to outdoors. How do I do it?

A1: Kitchen range hood or downdraft duct is generally smooth metal duct that is sized to match the outlet of the ventilation device. It is often six inch or seven inch round duct or the range hood may have a rectangular discharge. If it is rectangular, the fan will typically have a rectangular-to-round

adapter included. Always use a terminal device on the roof or wall that is sized to be at least as large as the duct. See Section 7 below for duct sizing requirements.

Q2: How do I know what the requirements are in my area?

A2: Ask your building department for that information. Some jurisdictions will accept metal flex, some will not.

5.2.1 Control and Operation.

The choice of control is left to the designer. It can be an automatic control like an occupancy sensor or a manual switch. Some products have multiple speeds and some switches have a delayed shutoff function that continues the exhaust fan flow for a set time after the occupant leaves the bathroom. In some cases, humidistats are used to automatically run exhaust fans based on local humidity levels. New control strategies continue to come to the market. The only restriction on controls is that they must not impede occupant control.

5.2.2 Ventilation Rate.

Section 5.2.2 simply requires the ventilation rate of intermittently operating local exhaust to meet the airflow rate shown in Table 5.1, which is reproduced below.

The kitchen exhaust requirement can be met with either a ceiling or wall mounted exhaust fan or with a vented range hood. Recirculating range hoods do not provide any ventilation and cannot be used to meet the requirements of the standard. If the flowrate is less than 5 air changes per hour, then a vented range hood (or integrated microwave/range hood unit) is required. See Section 5.3.2 below for a discussion of determining the kitchen volume.

ASHRAE ventilation standards have required a minimum intermittent flow of 100 cfm (50 L/s) for the range hood and 50 cfm (25 L/s) for the bath fan for over 25 years. The 100 cfm requirement for the range hood or microwave/hood combination is the minimum to adequately provide capture of the moisture and other products of cooking and/or combustion. Most range hoods provide more than one speed, with the high speed at 150 cfm or more – sometimes much more. Range hoods with high flow rates are often specified when large commercial-style stoves are installed. These systems can create significant pressure imbalances in the residence that can create numerous problems, including backdrafting of vented combustion appliances. Some codes, such as the International Residential Code (IRC 2009), require make-up air supply that is interlocked with the exhaust. See Section 6.4 for restrictions that may apply if large range hoods are installed.

Example 5-D Is an Intermittent Range Hood Required?

Q: I am building a house with a kitchen that is 12' x 14' with a 10 foot ceiling. What size ceiling fan is required?

A: The kitchen volume is 12' x 14' x 10' = 1680 cubic feet. 5 air changes is a flowrate of 1680 ft^3 * 5/ hr ÷ 60 min/hr = 140 cfm. So this kitchen must have a ceiling or wall fan of 140 cfm or a 100 cfm vented range hood.

TABLE 5.1 – Intermittent Local Ventilation Exhaust Airflow Rates

Application	Airflow	Notes
Kitchen	100 cfm (50 L/s)	Vented range hood (including appliance-range hood combinations) required if exhaust fan flow rate is less than 5 kitchen air changes per hour.
Bathroom	50 cfm (25 L/s)	

Section 5.3 Continuous Mechanical Exhaust

The standard allows the designer to install a local exhaust system that operates continuously and automatically. This is generally done when the local exhaust ventilation system is combined with a continuous whole-building ventilation system as required by Section 4.1 above. For example, if the whole-building exhaust is provided by a continuously operating exhaust fan located in the bathroom, this fan also satisfies the local exhaust requirement for the bathroom. The continuous local exhaust may also be part of the continuous whole-building ventilation system, such as a pickup for a remote fan or HRV/ERV system.

5.3.1 Control and Operation

All continuous fans, both local exhaust and whole-building systems, must be designed to operate automatically all the time. This is a 24/7 requirement and like the continuous whole-building controls required in Section 4.3 above, the occupant must have a readily accessible way of overriding or turning off the system. However, controls commonly used to operate intermittent fans such as occupant sensors, humidity sensors, or shut-off timers cannot be used to operate continuous exhaust fans.

As with the whole-building ventilation system, the standard requires that local exhaust systems be installed, but the occupant has the option of choosing not to operate the system.

5.3.2 Ventilation Rate

Section 5.3.2 simply requires the ventilation rate of continuously operating local exhausts to meet the airflow rate shown in Table 5.2, which is reproduced below.

TABLE 5.2 – Continuous Local Ventilation Exhaust Airflow Rates

Application	Airflow	Notes
Kitchen	5 air changes per hour	Based on kitchen volume
Bathroom	20 cfm (10 L/s)	

The requirement that a continuous kitchen exhaust have a capacity to provide 5 air changes per hour is due to the difficulty of a non-hood exhaust to adequately capture and remove contaminants released during cooking from kitchens which may be quite large, have an open-plan design, or have high ceilings. The standard does not dictate how the volume of the kitchen is calculated. It is up to the designer to decide what the kitchen volume is and to decide if a continuous approach is appropriate. Local building codes may have specific definitions of rooms.

5.4 Airflow Rating

In order to assure that the airflow required in Section 5 will actually be delivered, the airflow of the installed fan must be measured. This measurement must be made using a flow hood, flow grid, or other airflow measuring device. This requirement is identical to that in Section 4.3 for whole-building ventilation systems. However, for the local exhaust equipment, an alternative method is also allowed.

The alternate approach is to install a fan which has an airflow rating that meets or exceeds the required airflow when measured at 0.25" w.g. In addition, the duct system must meet the requirements of Table 5.3 in the standard, which is reproduced below. (Note that some of the table values shown in the standard were erroneous, as listed in errata issued by ASHRAE on July 21, 2010. The table below includes the correct values.) The certified airflow rating of a ventilation device is generally available from the manufacturer. It is also available for hundreds of products in the HVI Certified Products Directory at the HVI website (www.hvi.org). Manufacturers can choose whether to provide the certified data for posting at the HVI website, but all of them should have available the rated data at 0.25 inches of water column static pressure.

Example 5-E Continuous Kitchen Exhaust

Q1: The kitchen in an apartment is 5 ft. by 10 ft., with an 8 foot ceiling. If a continuous ceiling mounted exhaust fan is used, what must the airflow be?

A1: The kitchen volume is 5' x 10' x 8' = 400 cu. ft. 5 air changes equates to 400 ft3 x 5/hr ÷ 60 min/hr = 34 cfm.

Q2: The kitchen in a new house opens into the family room. There is no wall between the rooms. Where do I measure the kitchen volume?

A2: It is up to the designer (and code official) to determine where the kitchen ends and the living room begins. However, basing the design on a larger volume would be a conservative choice.

Q3: A new house has an open-design 12'x18' ranch kitchen with 12 foot cathedral ceilings. What airflow rate will be required for a continuous exhaust fan?

A3: The kitchen volume is 12' x 18' x 12' = 2592 ft^3. The airflow required is 2592 ft^3 x 5/hr ÷ 60 min/hr = 216 cfm. It would seem that a vented range hood, operated intermittently, might be a better choice in a kitchen this size.

The historical airflow rating pressure for residential fans is 0.1" w.g. (inches of water column static pressure). This is equivalent to about four feet of four-inch flex duct and a standard roof jack or wall cap. Since most houses and apartments require a longer length of ducting to reach the outside of the building, a higher rating pressure is used by Standard 62.2. After finding that a majority of the installed 3" ducted fans did not produce anywhere near their rated flow, Washington State adopted a higher rating pressure of 0.25 inches of water column in 1989 to reflect a more reasonable length of 3" duct. Still, maintaining the length of duct below the levels recommended in Table 5.3 of the standard is recommended to achieve the fan's rated flow rate. Another option for ensuring that flow rates are met is installing exhaust fans with larger duct diameters. The larger the duct diameter, the less significant the pressure rating in achieving the fan's rated airflow. For example, an exhaust fan with a 6" flex duct rated at 50 cfm at 0.1" w.g. may achieve its rated flow rate with duct lengths of over 250 feet.

The fans are tested in a qualified laboratory using a calibrated air flow test chamber. As with the sound testing, the fans are all mounted in a prescribed manner and the airflow is measured at ten levels of static pressure, producing a performance curve for the device from zero static to the static pressure which is high enough to stop airflow. Section 7.1 describes test standards that can be used to meet this requirement.

TABLE 5.3
Prescriptive Duct Sizing – IP Units

Duct Type	Flex Duct				Smooth Duct			
Fan Rating CFM @ 0.25 in. wg (L/s @ 62.5 Pa)	50 (25)	80 (40)	100 (50)	125 (65)	50 (25)	80 (40)	100 (50)	125 (65)
Diameter in. (mm)	Maximum Length ft. (m)							
3 (75)	X	X	X	X	5 (2)	X	X	X
4 (100)	70 (21)	3 (1)	X	X	105 (32)	35 (11)	5 (2)	X
5 (125)	NL	70 (21)	35 (11)	20 (7)	NL	135 (42)	85 (26)	55 (17)
6 (150)	NL	NL	135 (42)	95 (29)	NL	NL	NL	145 (45)
7 (175) and above	NL	NL	NL	NL	NL	NL	NL	NL

This table assumes no elbows. Deduct 15 feet (5 m) of allowable duct length for each elbow.
NL = no limit on duct length of this size.
X = not allowed, any length of duct of this size with assumed turns and fitting will exceed the rated pressure drop.

The second part of the prescriptive approach to system design is the use of minimum duct sizes as shown in Table 5.3. Using 0.25" w.g. of static pressure for the fan and tables from Manual D, the committee developed a table of minimum duct sizes and maximum duct lengths at various design airflows as a guide for the designer and the inspector. As can be seen, the higher the flow, the larger in diameter, or shorter in length the duct has to be. Also note that smooth duct can be used to manage longer duct runs. The table only gives a limited number of airflow ratings, but the designer and/or installer may interpolate between columns for different airflow values.

Example 5-F Prescriptive Duct Sizing

Q1: From Table 5.2, I have found that I need to provide 75 cfm of continuous ventilation to provide 5 ach to a kitchen. I plan to connect the fan to a roof vent termination using flex duct. The duct will be about 8 feet long, with no real elbows, but some slight bends in the duct. What size duct do I need to use?

A1: From Table 5.3, using the 80 cfm, flex duct column, we find that the maximum length with 4" duct is 3 feet, so you cannot use 4" duct. With 5" duct the maximum length is 70 feet, so that will clearly be adequate. Even if the bend in the duct is treated as an elbow, the allowable length only drops to 55 feet, more than adequate for the 8 feet required.

Q2: For the situation in Question 1, again providing 75 cfm, what size duct would I need if smooth metal duct were used? In this case the total length would increase to about 10 feet, and there would be 2 elbows.

A2: Using the 80 cfm, smooth duct column of Table 5.3, we find that the maximum length of 4" duct is 35 feet. Subtracting 15 feet for each of the 2 elbows leaves us with 5 feet, which is not long enough. But 80 cfm is a higher airflow than we need, so we are allowed to interpolate between the 80 cfm and 50 cfm columns. At 50 cfm, 4" duct can be 105 feet. Interpolating to 75 cfm we get (75-50)/(80-50)x(35-105)+105= 46.7 feet, rounded down to 46 feet. Subtracting 30 feet for the two elbows, we get 16 feet, which is more than the 10 feet needed, so 4" smooth metal duct can be used.

Q3: I will need a 100 cfm range hood. I have two possible duct routings. One is 15 feet long and will require 3 elbows. The other is 35 feet long but only requires one elbow. What size flex duct do I need to use?

A3: First, let's take the two routings and add in the correction for the elbows. Elbow corrections can be either added to the desired length or subtracted from the allowable length. In this case, we know the desired length, so we'll add the elbows. We get 15 feet plus 3 times 15 feet for a total of 60 feet, or 35 feet plus 15 feet equals 50 feet.

Looking at Table 7.1a, in the 100 cfm, flex duct column, we find that the maximum length with 5" duct is 35 feet, which is less than the adjusted length for either routing. With 6" duct, the maximum length is 125 feet, longer than either adjusted length. 6" duct would need to be used for either routing. Note: the building code may not allow flex duct to be used for the range hood, in which case smooth duct would be required. For smooth duct, 5" would be acceptable.

Section 6 Other Requirements

6.1 Adjacent Spaces

The standard requires that the air used for ventilation purposes comes from the outdoors. Air may not be drawn in as transfer air from other spaces that are outside the occupiable space of the dwelling unit. This is to prevent airborne pollutants originating in those other spaces from contaminating the living area. For example, drawing ventilation air from the garage could introduce VOCs, or pesticides into the indoor air. Drawing ventilation air from an unconditioned crawlspace could cause elevated allergen concentrations in the dwelling such as mold spores, or insect or rodent allergens. Likewise, drawing air from an adjacent dwelling could introduce unwanted contaminants such as cooking products or cigarette smoke into the unit.

In addition to designing the ventilation system to draw air from the outdoors, the standard also requires that measures be taken to prevent air movement between adjacent dwelling units and between the dwelling unit and other adjacent spaces, such as garages. Section 6.5 of the standard provides airtightness requirements related to garages and duct systems. Section 6.1.1 requires that doors between dwelling units and common hallways must prevent air movement by being substantially airtight through the use of gaskets or weatherstripping. This requirement does not apply to doors to outdoors. It also does not apply if the ventilation system explicitly requires transfer air from the corridor into the dwelling unit. It should be pointed out that the wording of the exception does not allow for transfer air from the dwelling unit into the corridor. A system which supplies ventilation air to the dwelling unit with exhaust via the corridor is not allowed.

6.2 Instructions and Labeling

The writers of the standard committee decided early on that with the requirement for mechanical ventilation came a responsibility to provide the occupant and/or owner with information on what the ventilation system was supposed to do and how to operate it. There has been a history of ventilation systems that worked initially but failed due to lack of information for the occupant or lack of maintenance. So the standard requires that the installer or builder provide written information on the basic ventilation concept being used and the expected performance of the system. These instructions must include how to operate the system and what maintenance is required. Chapter 13 of ASHRAE Guideline 24-2008 provides some guidance on what to provide and a suggested worksheet to list the details.

Because the concept of a designed whole-building ventilation system may be new to a lot of occupants, the standard requires that ventilation system controls be labeled as to their function. No specific wording is mandated, but the wording needs to make clear what the control is for and the importance of operating the system. This may be as simple as "Ventilation Control" or might include wording such as "Operate whenever the house is in use" or "Keep on except when gone over 7 days". If the system is designed to operate with a timer as an intermittent system, the labeling may need to be more complex. One option that has worked well in the Pacific Northwest is to affix a label to the electrical panel that provides some basic system operation information.

6.3 Clothes Dryers

The requirement of this section of the standard is very simple, "Clothes dryers shall be exhausted directly to the outdoors." This means that all laundry rooms must be built with a duct to the outdoors, designed to be connected to the dryer. Devices which allow the exhaust air to be diverted into the indoor space to provide extra heating are not permitted. This requirement is consistent with existing clothes dryer installation and design standards.

An exception to this requirement is for condensing clothes dryers, which collect the water from the exhaust air and then release the resulting airstream into the space. These units must be plumbed to a drain.

In multi-family buildings, multiple dryer exhaust ducts can be connected to a common exhaust only when the common exhaust duct meets the requirements of Section 7.3.

Example 6-A Clothes Dryer Exhaust Diverter

Q: I am building a home which has been purchased prior to completion. The buyer has asked for an exhaust air diverter to be installed in the dryer exhaust duct. He says that it is wasteful of heating energy to exhaust the warm humid air to the outdoors during the winter when the furnace and humidifier are working. He says that the screen on the diverter will prevent excess dust being released into the space. Can I install the device for him?

A: No. The writers of the standard have specifically prohibited the type of device your buyer wants to install. Significant amounts of dust are released from such devices, and the moisture in the dryer exhaust can lead to humidity problems as well.

6.4 Combustion and Solid-Fuel Burning Appliances

Background

The vent system of a combustion appliance is designed to safely carry the products of combustion out of the home. Typical vented combustion appliances include furnaces, boilers, water heaters, gas or wood fireplaces, and wood stoves. Depending on the fuel and operating conditions of the appliance, the makeup of the products of combustion will vary.

The vent system carries the products of combustion out of the home. The movement of air through the vent system relies on a pressure difference between the indoor and outdoor ends of the vent. When the pressure at the indoor end of the vent is higher than at the outdoor end, then the products of combustion will move out of the house. This pressure difference can be established in one of two ways, by mechanical means or through natural draft.

Many newer appliances include a combustion fan or forced draft blower which will provide the positive pressure at the indoor end of the vent. Other appliances, including most water heaters and solid-fuel appliances, rely on the buoyancy of the hot combustion gas to provide the driving force to move the gases through the vent. These are atmospherically vented (or natural draft) appliances.

Direct vent appliances have two air paths to the outdoors, one for air intake and one for venting combustion products to the outside. These appliances do not have air openings to the indoors, providing them with a high degree of isolation from the indoor conditions. These appliances are often referred to as "sealed combustion" although they are not necessarily completely sealed from the indoors.

Under certain conditions, flow through the vent system of any appliance, mechanically or atmospherically vented, can become reversed. This is commonly referred to as backdrafting, and it can have a number of causes. One generally benign type of backdrafting is referred to as spillage. This occurs when the appliance first starts. If the vent system is cold, there may be a short period of time before proper flow through the vent is established. As the vent warms up, proper flow will be established, and the spilled combustion gases will be drawn back up the vent through the draft diverter. This type of spillage is specifically allowed under product design certification standards for this warm up period, normally five minutes after burner ignition.

Other causes of backdrafting are more problematic, because if the proper flow through the vent is not established, then the appliance will continue to vent into the house. One of the primary combustion products is water, and continued backdrafting can lead to very high indoor humidity, with condensation and mold growth. Other components of combustion gases can be more dangerous, although they will not always be present in high quantities. The most significant of these is carbon monoxide, which can be fatal in a short time period if concentrations are high enough. Combustion products from oil and solid fuel appliances will contain other toxics. In any case, sustained backdrafting is to be avoided.

Proper appliance installation according to the code or standard and the manufacturer's instructions should provide a good margin of safety for protection against backdrafting. Many times, backdrafting is the result of low pressures inside the house, at the indoor end of the vent, causing vent airflow to reverse. A properly installed appliance is able to overcome some degree of depressurization, depending on the specifics of the appliance, vent system and weather conditions. Mechanically vented appliances are able to overcome significantly more depressurization than atmospherically vented appliances. A depressurization which exceeds that maximum depressurization of the appliance and vent system, whatever it is, will result in backdrafting.

One common source of depressurization which may exceed the capability of atmospherically vented appliances is an oversized kitchen exhaust fan. These are commonly installed with "commercial" styled appliances, either as a separate range hood or as part of the range itself. Downdraft ranges also tend to have large flow capacities, needed to provide adequate particle capture in the downdraft configuration.

The standard requires that the vent system be properly installed, as specified by the instructions from the appliance manufacturer and by specific model code documents. Often, the requirements of the manufacturer's instructions are based on the model code. There are three different standards or codes which set minimum requirements for design and installation of the vent system, depending on the fuel used – natural gas or propane, oil, or solid fuel. These are shown below in Table 6.1.

TABLE 6.1
Standards for Installation and Vent System Design for Combustion Appliances

Fuel	Applicable Code or Standard
Natural gas or propane	National Fuel Gas Code (NFPA 54-2002/ANSI Z223.1-2002)
Oil	Standard for the Installation of Oil-Burning Equipment (NFPA 31-2006)
Solid Fuel	Standard for Chimneys, Fireplaces, Vents and Solid-Fuel Burning Appliances (NFPA 211-2006)

All three are available for purchase from the National Fire Protection Association at http://www.nfpa.org/catalog/. The National Fuel Gas Code is also available from the American Gas Association at http://www.aga.org/Template.cfm?Section=Publications_Catalog&Template=/Ecommerce/EcomDefault.cfm. Note that these three standards are updated periodically. You should use the latest version of the applicable document, unless the building code in your area references some earlier version. ASHRAE will update Standard 62.2 periodically to reference the latest version, but there will be some lag time between the release of a new version of the NFPA document and when it is referenced in 62.2.

The standard also allows the use of other equivalent codes published by other organizations if the alternate is acceptable to the local building official.

The code or standard and the manufacturer's installation instructions will provide requirements for two aspects of the installation which are important from an indoor air quality standpoint – venting and combustion air. Other requirements are also included, such as fuel piping, appliance installation requirements, and procedures for placing the appliance into service.

Compliance with the venting requirements will involve determining the type of vent material to be used, the sizing of the vent system, and vent routing requirements.

Section 6.4 of Standard 62.2 includes a provision intended to prevent this type of backdrafting. If one or more large exhaust fans are installed in a home with atmospherically vented or solid fuel appliances, then the net exhaust flow must be reduced. The requirement is based on the two largest exhaust fans when they have a combined capacity that exceeds 15 cfm/100 ft² of floor area. This is 300 cfm in a 2,000 ft² house, or 450 cfm in a 3,000 ft² house. Many large range hoods have capacities which are 1,000 cfm or more. This provision applies only when the atmospherically vented appliance is inside the pressure boundary of the house, and does not include a summer cooling fan which is designed to be operated with the windows open. The standard specifically states that direct-vent appliances are not considered "atmospherically vented." (See Chapter 3 of the standard for a definition of pressure boundary.)

The net exhaust flow can be reduced in one of two ways. First, the speed and flowrate of one or both of the fans can be reduced so that the combined flow is less than 15 cfm/100 ft². This can also be solved by installing a supply fan to balance the flow. This supply fan must have a capacity that is greater than 15 cfm/100 ft² less than the capacity of the exhaust fans. In other words:

Supply Capacity ≥ Exhaust Capacity – 15 cfm/100 ft² x Floor Area

This supply fan must have interlocked controls such that whenever the largest exhaust fan is turned on, the supply fan comes on as well.

6.5 Airtightness Requirements

Garages often contain numerous sources of contaminants. These include gasoline and exhaust from vehicles, pesticides, paints and solvents, etc. Air flowing from the garage into occupied space can carry these contaminants into the dwelling, resulting in poor indoor air quality. Air handling units located in the garage can be a particular problem when there are leaks in the return side of the unit or in the return ducts pulling contaminated air into the dwelling. Because of the larger pressure differences across these leaks, compared to leaks between the garage and the dwelling, significantly larger contaminant levels can result.

6.5.1 Garages

The standard requires that when garages are attached to the house, contaminants must be prevented from entering the house. This section specifies air sealing requirements for any wall, ceiling or floor assemblies separating occupiable space from a garage. Specifically, all:

- Joints,
- Seams,
- Penetrations,

Example 6-B Large Exhaust Fan

Q1: I am building a custom home with 4 bedrooms and 3,600 sq. ft. The kitchen will have a high end range hood which has three speeds, nominally 1000 cfm, 1400 cfm and 1600 cfm. The house will be heated with a gas furnace located in the basement. If I am using a central exhaust fan for the whole-building ventilation of 90 cfm, and there is a clothes dryer installed, how large must my compensating supply fan be?

A1: You must use the high speed value for the range hood of 1600 cfm. The clothes dryer will have a flow that is assumed to be 150 cfm for sizing purposes. These two flows are added together for a total exhaust capacity of 1750 cfm. Since the whole-building ventilation fan is not one of the two largest exhaust fans, it does not figure into sizing the supply fan. Using the equation above, the supply fan must be at least 1750 cfm – 15 cfm x 3600 ft² / 100 ft² = 1210 cfm.

Q2: The same custom house will have the furnace located in the garage instead of the basement. Does that change anything?

A2: The garage and the attic would both normally be considered outside the pressure boundary, so no makeup air would be required. An exception to this would be if the attic is specially designed to be inside the pressure boundary, then the answer would be the same as for Q1.

Q3: For this house, I need to keep the furnace in the basement. What are my options that would avoid using the compensating supply fan?

A3: There are several things you could do. First, you could use direct vent appliances which would give higher efficiency and would not require a supply fan. Use of electric appliances would not require a supply fan (but might have higher energy costs). You could use a lower capacity range hood, one that is less than 390 cfm (15 cfm x 3600 ft² / 100 ft² – 150 cfm). Use of supply only whole-building ventilation would allow the hood capacity to increase to 480 cfm (15 cfm x 3600 ft² / 100 ft² – 150 cfm + 90 cfm). There are also range hoods available in the commercial market which have integrated supply fans, and one of these units would be acceptable too.

- Openings between door assemblies and their respective jambs and framing, and
- Other sources of air leakage through the assembly,

must be sealed by being:

- Caulked,
- Gasketed,
- Weather-stripped,
- Wrapped, or
- Otherwise sealed to limit air movement.

In addition, doors must be weatherstripped or gasketed.

6.5.2 Space-Conditioning System Ducts

It is not allowed for air handling systems which serve occupiable space to supply air to the garage or return air from the garage. If space conditioning is provided to the garage, it must be a separate system.

If ducts are located outside the pressure boundary (including in the garage, attic or a crawl space), the system must be sealed to obtain a total system leakage of no more than 6% of the total fan flow when measured at 0.1 in. of water (25 Pascals). Leakage tests must be conducted in accordance with California Title 24 (2001) or equivalent. Method D of ANSI/ASTM E1554 is specifically cited as an equivalent test procedure. This is especially important if an air handling unit or return ducts are located in the garage.

6.5.3 Exception

This section must be disregarded. It appears only through an error in the production process, and has been removed by an errata posted on the ASHRAE website on July 21, 2010.

Example 6-C Garages

Q1: The building designer located the air handler in the garage. The air handler installed has an un-ducted return. Is this acceptable?

A1: No, this design is not allowed because an un-ducted return would draw all return air from the garage. Typically, building codes would not allow this design either.

Q2: The building designer located the air handler in the garage. The main return trunk from the dwelling is connected to the air handler. Is this acceptable?

A2: Yes, provided that the duct system is leak tested at 25 Pa. in accordance with California Title 24 (2001) or equivalent, and sealed, if necessary, to have leakage no greater than 6% of the total fan flow.

Q3: The building designer located the air handler in the dwelling unit. A return duct runs through the garage to a bedroom above the garage. The duct has only 4 feet of length in the garage. How do I test that length of the duct?

A3: This design is allowed but the entire duct system must be leak tested at 25 Pa. in accordance with California Title 24 (2001) or equivalent., and sealed, if necessary, to have leakage no greater than 6% of the total fan flow. There is no test available to leak test only the garage portion of the duct system. Visual inspection of the duct section in the garage, using leak-identifying smoke if possible, is recommended, but is not a requirement of the standard.

6.6 Ventilation Opening Area

> *Background*
>
> *The whole-building ventilation rates provided in Section 4.1 are intended to provide adequate ventilation to typical new dwellings under normal circumstances. On occasion, however, houses experience unusual circumstances where high levels of contaminants are released into the space. When this occurs, some means of providing the increased levels of ventilation required to remove the contaminants is needed. Openable windows are the most likely means of providing the additional ventilation. The requirements in Section 6.6 are based on typical window requirements in the building codes.*

This section of the standard requires ventilation openings in habitable spaces, toilets and utility rooms. Ventilation openings usually will mean openable windows, although a dedicated non-window opening for ventilation is acceptable. Spaces that meet the local exhaust requirements of Section 5 are exempted from this requirement because they have mechanical ventilation.

6.6.1 Habitable Spaces

Habitable spaces are required to have ventilation openings with openable area equal to at least 4% of the space floor area (but not less than 5 square feet). Rooms people occupy are considered habitable space. Dining rooms, living rooms, family rooms, bedrooms and kitchens are considered habitable space. Closets, crawl spaces, garages and utility rooms are generally not. If the washer and dryer are located in an open basement that is also the family room, it would be considered habitable space.

Of course the openings do not have to be provided by windows. They can also be provided by openable, insulated, weather-stripped panels.

6.6.2 Toilets and Utility Rooms

This section extends the requirement for ventilation opening areas to toilets and utility rooms. The openable area required is still at least 4% of the space floor area, but the minimum area required is 1.5 square feet rather than 5 square feet. There are two exceptions to 6.6.2. If a ducted dryer exhaust is used in a utility room (as required by Section 6.3), the requirement for ventilation openings does

Example 6-D Ventilation Openings

Q1: I am building a house with a 14 ft. by 12 ft. bedroom What size window do I need to install?

A1: It depends on the type of window. The standard requires that the openable area of the window, not the window unit, be 4% of the floor area, or 14' x 12' x 0.04 = 6.7 ft^2. The fully opened area of the window or windows must be greater than 6.7 square feet. The requirement for this example can

be met using two double hung windows each with a fully opened area of 3.35 square feet. Any combination of windows whose opened areas add up to at least 6.7 square feet will meet the requirement.

Q2: The same house has a bedroom which is 10 ft. by 10 ft. What size opening in the window do I need for this room?

A2: Based on the floor area, the required ventilation opening area is 10' x 10' x 0.04 = 4 ft^2. However this is less than the 5 sq. ft. lower limit, so a window with an opening of 5 square feet of venting area is required.

6.7 Minimum Filtration

Background

MERV 6 efficiency filters are required by the standard because field research[1] *has demonstrated that less than this level of efficiency does not significantly deter the accumulation of particulate contamination on the HVAC heat exchange equipment components and related distribution system surfaces over the lifetime of the equipment. Such accumulation can impair the operating efficiency of the system resulting in excessive energy usage and it can become a potential reservoir for odor and moisture, as well as a nutrient source for fungal growth. Improved filtration also lessens the amount of recirculated particulate matter in the home, keeping the space cleaner of nuisance housekeeping dust, as well as reducing respirable particles that are potentially harmful or irritating to occupants--resulting in improved indoor air quality. MERV 6 is the required minimum efficiency level. "MERV" stands for Minimum Efficiency Reporting Value and the "6" is a performance level based on ASHRAE Standard 52.2-2007*[2].

The limitation of the filtration requirement to systems with 10' of ductwork is to avoid undue requirements in units having little distribution ductwork and having neither the space nor the airflow capacity for better filters, such as window units.

The location and installation of the filter is important in order to assure that 100% of the recirculated air is treated and that the filter cartridge can be easily accessed by the home owner for timely inspection and change-out.

In addition to the minimum filtration efficiency rating, the selection and sizing of the filter is critical to avoid excessive air restriction and related harm to the HVAC equipment.

The standard requires that particulate air filtration of no less than MERV 6 efficiency be installed in any HVAC system having more than 10 ft of ductwork. The particulate filter must be installed such that all of the air circulated through the furnace or air handler is filtered prior to passing through the thermal conditioning portion of the system. In addition, the standard requires that the filter be located and installed for easy access and service by the homeowner.

MERV 6 filters are normally 1" thick with a pleated media configuration to attain the proper efficiency and airflow performance. Allowable filters will be labeled with a MERV 6 designation. Note that electrostatic air cleaners do not meet the requirements of the standard as none of them are evaluated to carry a MERV designation. However, they may be installed downstream of a MERV 6 prefilter.

The filter retainer section must be easily accessible by the homeowner to assure continued monitoring and replacement. Figure 6.1 demonstrates the typical location options of filter installation. The filter bank may be located in the air handler/furnace (1); in the return air plenum near the air

[1] Burroughs, H. E. *Role of Filtration Efficiency in HVAC System Cleanliness and Building Security.* American Filtration and Separations Society, 9th World Filtration Congress, New Orleans, 2004

[2] ASHRAE Standard 52.2-2007 *Method of Testing General Ventilation Air-cleaning Devices for Removal Efficiency by Particle Size,* American Society of Heating Refrigerating, and Air Conditioning Engineers, Inc., 1791 Tullie Circle NE, Atlanta GA 30329.

handler (2a); in the return air plenum with a deep pleat cartridge (2b); angled across the return air plenum to enhance cross-section (3); or situated in a wall return grille (4).

The MERV 6 pleated filter provides enhanced particulate arrestance, but also provides longer service life than the conventional low efficiency flat panel filter. Typically, the pleated type filter will last three months or longer because of its increased surface area, depending upon operating conditions, as compared to the typical one-month life cycle of disposable fiberglass filters. The deeper pleated versions will typically provide even longer life cycles, up to one year or more. To assist the homeowner to properly monitor and maintain the filter system as required in Section 6.2, a service label such as the example shown in Figure 6.2 can be applied to the ductwork near the filter rack. Such a label is not required by the standard, but may be helpful to the occupant to get the best performance from the filtration system.

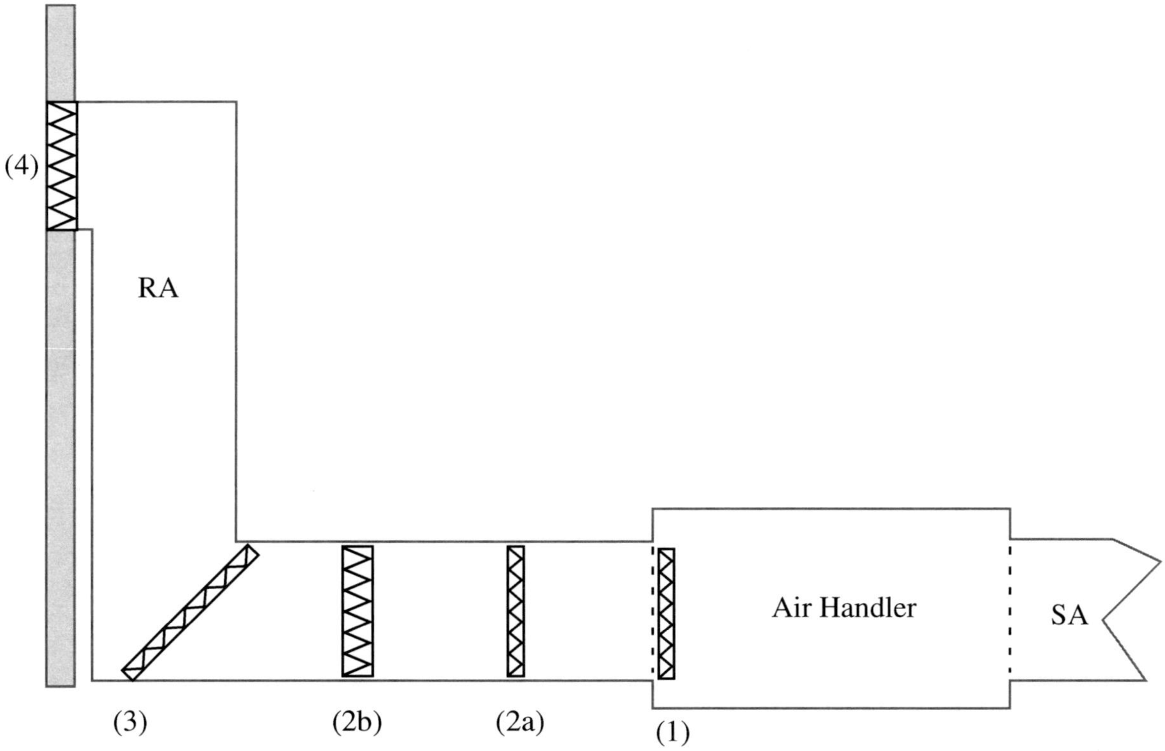

Figure 6.1 Filter Location Options.

FILTER SERVICE RECORD		
For Filter Service or Replacement Cartridges, Contact:		
Installation Date	**Related Pressure Drop**	**Service/Monitor Date**

Figure 6.2 Example Filter Label for Application on Filter Housing

6.8 Air Inlets

When the ventilation system is designed with air inlets, the inlets must be located away from locations that can be expected to be sources of contamination. The minimum separation is 10 feet (3 meters). Inlets include not only inlets to ducts, but windows which are needed to reach the opening area specified in Section 6.6. (If a room has more windows than is required by Section 6.6, the additional windows can be closer than the 10 foot (3 meter) limit.

The standard lists some likely sources of contaminants. These include: a stack, vent, exhaust hood or vehicle exhaust. For typical residential applications, the sources will include:

- vents from combustion appliances,
- chimneys,
- exhaust fan outlets,
- barbeque grills,
- locations where vehicles may be idling for any significant length of time, and
- any other locations where contaminants will be generated.

The standard also requires that air intakes be placed so that they will not become obstructed by snow, plants, or other material. Forced air inlets must also be equipped with insect/rodent screens, where the mesh is no larger than 1/2 inch (13 mm).

There are three exceptions to the separation requirements.

- Windows or ventilation openings in the wall can be as close as three feet to sources of contamination which exit through the roof or to dryer exhausts.
- There is no minimum distance between windows and the outlet of a local exhaust outlet from kitchens or bathrooms.
- Vent terminations which meet the requirements of the National Fuel Gas Code[6], which has its own separation and location requirements, do not need to meet the requirements of Section 6.8.

6.8.1 Ventilation Openings

Ventilation openings, which include windows, skylights, through-the-wall inlets, window air inlets, or similar devices, shall be readily accessible to the occupant. This means that the occupant must be able to operate the opening without having to climb on anything. An openable skylight, which is used to meet the ventilation opening requirements of Section 6.6, must have some means of being operated while standing on the floor – a push rod, a long crank handle, or an electric motor.

If a ventilation opening is covered with louvers or otherwise obstructed, the openable area used in meeting the Section 6.6 requirements is the unobstructed free area through the opening.

Example 6-F Ventilation Opening Louvers

Q: There are fixed wooden louvers over a window in a bedroom. The louvers have slats that are 1/8 in thick, and they are spaced 1 in. apart. What is the reduction in openable area?

A: Assuming that the 1 inch spacing was measured perpendicular to the slats (the correct way), then the reduction is the slat thickness divided by the spacing, or 1/8. So the credited opening area is the original opening area x (1" – 1/8")/1" = 7/8 of the original opening area.

Section 7 Air-Moving Equipment

7.1 Selection and Installation

Background

Ventilation systems often do not provide the level of ventilation that was intended. Occupants sometimes choose not to use (or even disable) the ventilation system because they are too loud. Duct systems impose higher static pressure than what is assumed at the typical rating point, so the airflow provided is lower than expected. Particular fans or devices may not always achieve rated performance.

To avoid these problems, the standard requires that the products be tested using nationally-recognized standards for both airflow and sound. The test results are required to be certified by a third-party organization, not just by the manufacturer. Devices installed to meet the standard must then meet certain criteria using the validated test results.

One organization that provides certification is the Home Ventilating Institute (HVI) (www.hvi.org), but it is only an example. HVI is no longer a division of AMCA, but continues to serve the residential market while AMCA serves the commercial and industrial market. Other organizations may also have residential product performance certification programs. The cited test standards are nationally-recognized as the leading methods of test for airflow and sound. HVI has test procedures for testing residential ventilation products using these standards that are available at the HVI website and the organization maintains a certification program for both sound and airflow performance ratings.

This section of the standard requires that equipment used to comply with the standard is selected based on tested and certified ratings of performance for airflow and sound. The standard lists five standards[3] [4] [5] [6] [7] which are acceptable, but equivalent other standards may also be used. When selecting fans for use in meeting the requirements of the standard, you must check to see that they have been tested and their performance meets the requirements of Section 7.2 and 7.3.

In addition, the standard requires that the fans be installed in accordance with the manufacturer's instructions. You must review the installation instructions and other literature shipped with the fan, and make sure that the installation complies with those instructions.

---

[3] *ANSI/ASHRAE Standard 51-1999/AMCA 210-99, Laboratory Methods of Testing Fans for Aerodynamic Performance Rating.* American Air Movement and Control Association International, Inc., Arlington Heights, IL.

[4] *ANSI/AMCA 300-05, Reverberant Room Method for Sound Testing of Fans.* American Air Movement and Control Association International, Inc., Arlington Heights, IL.

[5] *HVI 915-06, Procedure for Loudness Rating of Residential Fan Products,* Home Ventilating Institute, Arlington Heights, IL.

[6] *HVI 916-09. Air Flow Test Procedure.* Home Ventilating Institute, Arlington Heights, IL.

[7] *HVI 920-09, Product Performance Certification Procedure.* Home Ventilating Institute, Arlington Heights, IL.

7.2 Sound Ratings for Fans

One common reason for not using ventilation equipment, particularly local exhaust fans, is the noise they create. To address this, the standard requires that certain fans be rated for sound and have ratings below specified limits. The rating must be done at an airflow that is no less than the airflow the fan must provide to meet the standard. The sound limits are specified in two subsections.

There is an exception to the requirements for sound ratings. Fans which are mounted outside the habitable space and outside any bathroom, toilets or hallway, and have at least 4 ft (1 m) of ductwork between the fan and the intake grille, do not need to meet any sound rating requirements.

Because of the variables in length and type of duct and grille, there is no clearly repeatable way to specify a sound level for ventilation devices that are not mounted in the ceiling or wall surface. Consequently, air handlers, HRV/ERVs, inline fans and remote fans are exempted from the sound rating requirements that apply to surface-mounted fans. However, to reduce the amount of fan and/or motor noise that could come down the duct to the grille, the standard sets a minimum of four feet of ductwork between the grille and the ventilation device. This may still produce an undesirable amount of noise for the occupant, especially if hard metal duct is used. Flexible insulated duct or a sound attenuator to will reduce the transmitted sound into the space.

7.2.1 Whole-Building or Continuous Ventilation Fans

Background

There have been many complaints about the noise level of both bath fans and range hoods over the past 25 years. Utility programs began setting maximum sound levels for whole-building exhaust ventilation systems in 1984 when they first began including such systems. Initially, the sound level was set at 2.5 sones for fans that were set to operate for 2-4 hours a day. By 1986, this requirement had dropped to 1.5 sones and it dropped still further to 1.0 sone in 1990 when the operation requirement was raised to 6-8 hours a day. In 1991, the State of Washington began requiring whole-building exhaust fans rated at 1.5 sones or less and operated for at least 8 hours a day. The experience over the past twenty years is that the occupant is more willing to operate a fan when it is quieter. This standard requires that all continuous ventilation fans be rated and certified as producing no more than 1.0 sone. Fans are available that are rated at less than 0.3 sone.

Fans which are installed to meet continuous ventilation requirements must be rated at 1.0 sone or less. This applies to both whole-building ventilation fans and continuous local exhaust fans.

7.2.2 Intermittent Local Exhaust Fans

Fans which are installed to meet ventilation requirements as intermittent fans, whether the whole-building or local exhaust requirements, must be rated at 3.0 sone or less, unless the maximum rated airflow is greater than 400 cfm (200 l/s).

The standard extends the concept of quiet fans to range hoods and regular bath fans, not just the whole-building ventilation system, but it sets different limits. While the whole-building fan or other combined systems that operate continuously must be rated at 1.0 sone or less, the intermittent local exhaust fans must be rated at 3.0 sones or less at their rating point. So the bath fans must be rated at a maximum of 3.0 sones, somewhat quieter than the typical bath fan that would have a sone rating of 4.0 sones or higher. Range hoods must also be rated at 3.0 sones or less, but this is at their required "working speed" of 100 cfm or so. Most range hoods have maximum speeds of much more than 100 cfm, but 100 cfm is the minimum airflow that is required by the standard. A range hood that is rated at 8 sones at its highest speed of 600 cfm may be only 1.5 sones or less at its lowest "working speed" of 100 cfm. This data is available from the manufacturer and/or from the HVI website (www.hvi.org).

Background

Sound tests are performed in a reverberant sound room or chamber that is acoustically isolated from its surroundings. The fan is mounted in a manner that is required by the test procedure so that the test is the same for all fans of similar type. The sound measurements are taken with a specialized microphone and the data is sent to a calibrated sound analyzer computer. The background sound of the chamber is measured, as well as a reference sound source with known acoustic characteristics. The fan is run at its rating point and sound measurements are taken at 24 frequencies. Logarithmic data in decibels is then recalculated to sones, a linear measurement that reflects the human response to sound levels. The sone data at all the frequencies is then reduced to a single sone level that can be compared in a linear fashion. A 2.0 sone fan sounds twice as loud to the human ear as a 1.0 sone fan.

Figure 7.1 Fan Mounted for Sound Testing

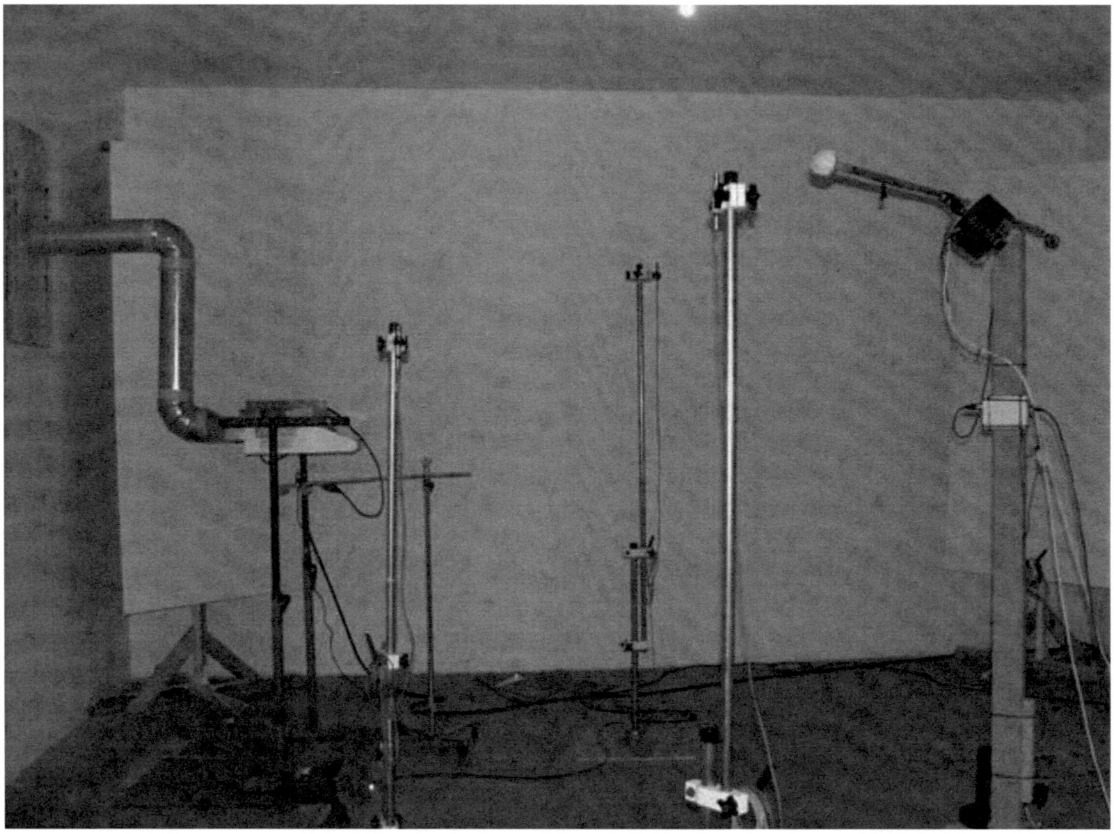

Figure 7.3 Microphone Array in Sound Room with Fan Mounted for Testing

Figure 7.3 A Fan Mounted to the Airflow Test Chamber at Texas A&M.

7.3 Multi-Branch Exhaust Ducting

The standard contains restrictions on several situations where multiple exhausts are connected through a combined duct system. These restrictions are intended to prevent one exhaust fan from sending air into a space served by a different fan.

The first restriction is that if more than one exhaust fan in a dwelling shares a common duct, then each fan must be equipped with a backdraft damper. Obviously, the system designer does not want to blow moisture or odors from one bathroom into another, so the standard addresses this with a requirement for backdraft dampers.

The other restriction applies to multiple dwelling units. Sometimes a single remote fan or HRV/ERV will exhaust from several locations in a house or unit or from several units in a multifamily building. See Figure 7.4, below. This section does not preclude the use of that type of system, but it does require that either the shared exhaust fan operate continuously so that air exhausted from one bathroom or unit is not allowed to go into another, or that backdraft dampers are used.

In multifamily buildings, fire codes may impose additional restrictions. There will generally be a requirement for a fire damper at the penetration of a rated shaft or a metal subduct that turns up a minimum of 22 inches plus a continuously operating fan at the top of the shaft. The 2003 International Building Code has an exception that allows the use of individual fans with radiant ceiling dampers discharging into dedicated metal ducts running to the roof without rated shafts or rooftop fans. This exception also allows attic-mounted multiport exhaust fans with pickups in various units that have a radiant ceiling damper behind the grille.

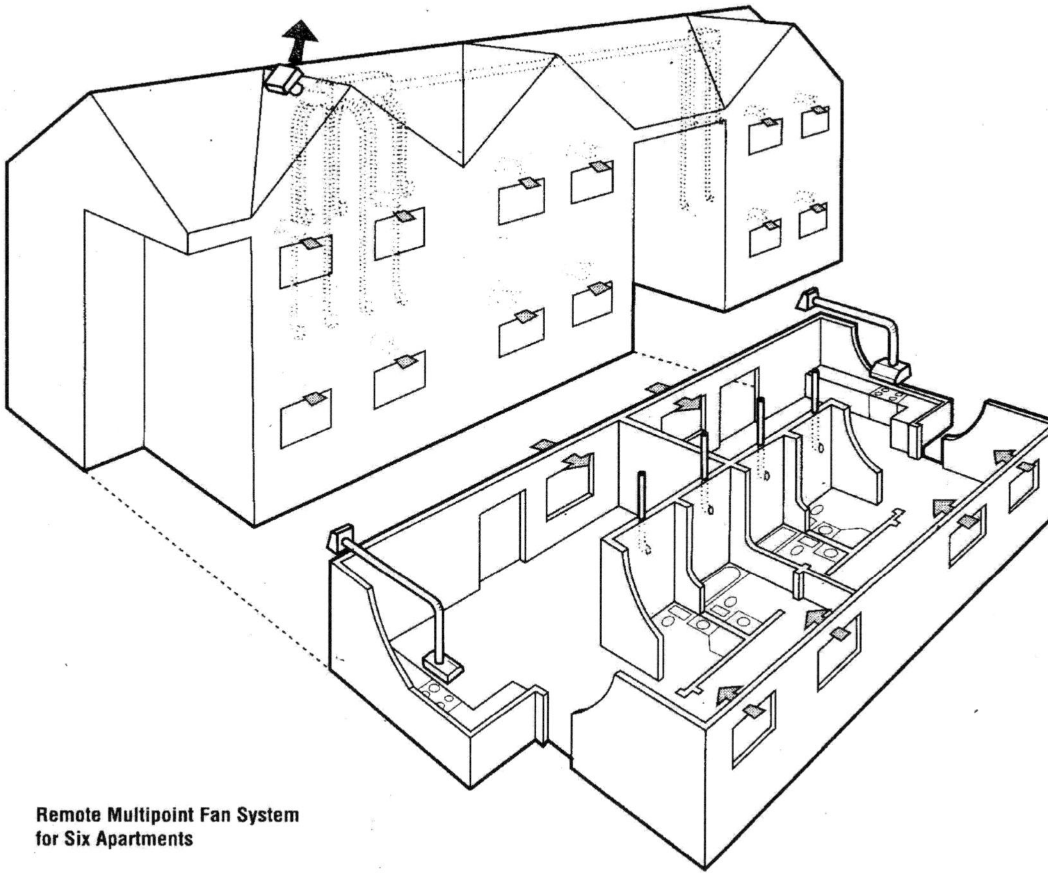

Figure 7.4 Diagram of a Remote Multipoint Exhaust Fan System

Section 8 –Climate Data

Section 8 presents climate data needed to determine where certain exceptions or requirements apply. Figure 8.1 in the standard presents a map of North America which shows the location of eight different climate types. The climate types are:

- Hot-Humid
- Hot-Dry
- Mixed-Dry
- Mixed-Humid
- Marine
- Cold
- Very Cold
- Subarctic/Arctic

Of the eight, two are used for requirements of the standard – Hot/Humid and Very Cold. The requirements for buildings in Very Cold climates are meant to apply to those in the Subarctic/Arctic climate as well. Figure 8.2 presents the climate zone map from the International Energy Conservation Code (IECC). Climate zones are numbered 1 through 8, and have moist, dry and marine subzones, lettered A, B and C, respectively. Exceptions in the standard apply to climate zones 1 and 2.

Table 8.1 in the standard lists a selection of U. S. cities which are located in the hot-humid climate zone. This is not an exhaustive list, but any nearby locations are likely in the same climate zone. Table 8.2 lists U.S. cities located in the very cold climate zone.

Section 9 – References

Section 9 of the standard simply lists 13 other standards that are used in Standard 62.2.

Appendix A – Existing Buildings

Background

Installation of local exhaust fans in bathrooms and kitchens during initial construction of a building can generally be done at reasonable cost. For existing buildings, however, installation of such fans can be prohibitively expensive. Similarly, upgrading existing exhaust fans that may not meet the requirements of the standard, may or may not be particularly expensive, but will often provide only limited benefit over the existing, but still functional fan.

There are numerous programs that would like to use Standard 62.2 when upgrading existing homes. However, some programs weatherizing existing homes found that upgrading these dwellings to be prohibitively expensive, particularly if installation of local exhaust fans was required. An alternative method for existing buildings to "meet Standard 62.2" was needed.

The alternative approach is described in Appendix A of the standard. A whole-building mechanical ventilation system will still be required in most cases, but if this system provides enough extra ventilation air, then the appendix allows exceptions to certain other requirements of the standard.

Section A1 Summary.

Appendix A of the standard provides an alternative compliance path intended to make it easier for existing buildings to comply with the standard. In general, the whole-building ventilation rate, as determined from Section 4.1 of the standard, is increased to compensate when it is impractical to provide kitchen or bathroom

Example A-A Determining Whether Use of Appendix A is Allowed

Q1: An electric utility company is sponsoring a home weatherization program which provides matching funding to homeowners who have an energy audit and invest in energy conservation measures identified by the audit. There are a number of required measures for the home to qualify for funding, one of which is that the home meets the requirements of Standard 62.2-2010. The utility has ruled that multifamily buildings may comply with the Standard using Appendix A procedures, but that single family buildings may not. Compliance with the standard is not required by local codes. Is this approach acceptable?

A1: In this case, the utility company becomes the authority having jurisdiction, and they have the right to specify the terms of their program.

Q2: A municipality has enacted regulations which require that new and significantly renovated residential buildings must comply with Standard 62.2-2010. The regulation specifies that significantly renovated existing buildings without local exhaust fans installed in bathrooms or kitchens may comply using Appendix A if those rooms have openable windows. If the bathrooms or kitchens have existing exhaust fans that do not comply with the requirements of the standard, then Appendix A may NOT be used for compliance, the existing fans must be upgraded. Is this acceptable?

A2: Section A1 of the standard is clear that the authority having jurisdiction may specify when and how Appendix A may be used for compliance. A local municipality is clearly an authority having jurisdiction, and they may impose restrictions on the use of Appendix A as they see fit.

exhaust equipment that complies with Section 5.

Existing buildings, as used in Appendix A, means buildings that have been occupied without meeting the requirements of the standard. Section A1 also specifies that it is up to the authority having jurisdiction to determine how and when compliance with the standard using Appendix A is allowed.

Finally, Section A1 makes it clear that while Appendix A specifies alternatives to compliance with specific sections in the body of the standard, the requirements in all other sections of the standard must still be met.

Section A2 Whole-Building Mechanical Ventilation Rate

This section specifies that the whole-building ventilation rate, as determined in Section 4.1, shall be adjusted. There are two adjustments that can be made. First, there is an increase in the ventilation rate that is determined in Section A3. Second, if the building airtightness has been measured, then Section 4.1.3 can be used to reduce the required mechanical ventilation rate. In other words, if the building is leaky, then the infiltration credit can be used to offset mechanical ventilation.

Section A3 Local Exhaust

Section A3 specifies how much the mechanical ventilation rate (from Section 4.1) must be increased when bath or kitchen exhaust systems do not meet the requirements of Section 5.1.

The first requirement of A3 is that if kitchen or bath exhaust equipment is being replaced or installed, then it must meet the requirements of Section 5.

The increased mechanical ventilation rate is determined using a three step process, with each of the steps being described in a separate subsection.

Section A3.1 Initial Room Airflow Deficit

The first step in the calculation is to compare the exhaust flow provided by the existing equipment, if any, to the requirement from Table 5.1. The measured or rated exhaust flow of the existing equipment to be used in this calculation must be determined following the requirements of Section A4.2. The Initial Room Airflow Deficit is simply the requirement for the room from Table 5.1 minus the existing equipment exhaust flow. It is best practice to measure the exhaust fan airflow of the existing equipment. If there is no exhaust device or if the existing device can neither be measured nor rated, the exhaust device airflow shall be assumed to be zero.

Section A3.2 Window Opening Credit

Step two in calculating the adjusted mechanical ventilation rate is to determine the Window Opening Credit. If the room (kitchen or bathroom) which has inadequate exhaust capacity has an openable window, then the Initial Room Airflow Deficit may be reduced by 20 cfm. This reduction is a fixed quantity, regardless of the size of the window. This credit may only be taken if the authority having jurisdiction approves openable windows as a permissible method for providing local exhaust.

Section A3.2 Required Additional Airflow

The final step in calculating the adjusted mechanical ventilation rate is to add the Initial Room Airflow Deficit for each kitchen or bathroom, subtract 20 cfm for each of those rooms with an openable window, then divide that total by four. This adjustment is then added to the total whole-building mechanical ventilation rate as determined in Section 4.1, including any infiltration credit adjustment as described in Section 4.1.3.

1. Add Initial Room Air Deficit for all applicable kitchens and bathrooms
2. Subtract 20 cfm for each kitchen and bathroom with an openable window
3. Divide the total by four. The result is the Required Additional Airflow.
4. Add to the whole-building mechanical ventilation rate determined in Section 4.1,
5. Deduct the infiltration credit determined in Section 4.1.3.

Section A4 Air Moving Equipment

Section A4 emphasizes that any air moving equipment that is being replaced or added must meet the requirements of Section 6 and Section 7.

Section A4.1 Selection, Installation and Sound Rating

This section specifies that the requirements of Sections 7.1 and 7.2, covering Selection and Installation and Sound Ratings do not need to be met for existing equipment in existing buildings.

Section A4.2 Airflow Rating

Section A4.2 describes two options for determining the airflow value for existing fans. This value can then be used to determine if it meets the requirements of Table 5.1 or, if not, in Section A3 to calculate the adjusted whole-building mechanical ventilation rate.

Example A-B Determining the Airflow Rating of an Existing Exhaust Fan

Q1: An existing home has a bathroom exhaust fan. The homeowner has retained the literature for the fan, and it is listed as having airflow of 50 cfm at 0.1" w.c. The ductwork connecting the fan to the outdoors is buried between floors. What airflow value can be used when determining the Initial Room Airflow Deficit in Section A3.1?

A1: In this case the rated airflow at 0.1" w.c. may be used if it is reduced by 25%. 25% of 50 cfm is 12.5 cfm, so the airflow value to be used in 50 cfm – 12.5 cfm = 37.5 cfm.

Q2: Another existing home has an exhaust fan in the bathrooms but no information on the fan is available. What airflow value can be used?

A2: In this case, the fan airflow must be measured using a flow hood, flow grid, or other airflow measuring device.

Q3: A kitchen range hood in an existing home is vented to the outdoors. The homeowner has retained a user's manual for the hood which describes the hood as having a two-speed fan with flow rates of 200 cfm and 500 cfm, but no test standard or test conditions are mentioned. Since both airflow rates are well above the kitchen exhaust rate of 100 cfm from Table 5.1, can the Initial Room Airflow Deficit for the kitchen be set to zero?

A3: Since the test conditions used to find the airflow rating for the hood of 200 cfm or 500 cfm are unknown, the airflow rate must be measured (using a flow hood, flow grid, or other airflow measuring device). If the measured flow rate of the hood exceeds the 100 cfm required by Table 5.1, then the Initial Room Airflow Deficit will be zero.

The preferred method for determining the airflow rate is by taking measurements using a flow hood, flow grid, or other airflow measuring device. The alternative method is to use the rated airflow for the fan at 0.10" w.c. (25 pa), reduced by 25%.

Example A-C Determining the Initial Room Airflow Deficit

Q: An existing home has two bathrooms and a kitchen. One bathroom has an exhaust fan which has a measured airflow of 37 cfm, but no openable windows. The other bathroom has no fan but does have an openable window. The kitchen has an exhaust fan with a measured airflow of 62 cfm, but no openable window. What is the Initial Room Airflow deficit for each room?

A: For the bathrooms, Table 5.1 requires an exhaust capacity of 50 cfm. 100 cfm is required for the kitchen. The Initial Room Airflow Deficit is the required airflow from Table 5.1 minus the airflow rating of the existing fan. For the first bathroom, this value is 13 cfm (50 cfm – 37 cfm). For the other bathroom, the Initial Room Airflow Deficit is 50 cfm (50 – 0). For the kitchen it is 38 cfm (100 – 62). The openable window in bathroom two is not included in determining the Initial Room Airflow Deficit, but will be used later for the Window Opening Credit.

Example A-D Determining the Required Additional Airflow

Q: An existing home has two bathrooms and a kitchen. One bathroom has an exhaust fan which has a measured airflow of 37 cfm, but no openable window. The other bathroom has no fan but does have an openable window. The kitchen has an exhaust fan with a measured airflow of 126 cfm, but no openable window. What is the Required Additional Airflow for the home?

A: Calculating the Required Additional Airflow is a three step process. Step 1 is to determine the Initial Room Airflow Deficit for each room. For the bathrooms, Table 5.1 requires an exhaust capacity of 50 cfm and 100 cfm is required for the kitchen. The Initial Room Airflow Deficit is the required airflow from Table 5.1 minus the airflow rating of the existing fan. For the first bathroom, this value is 13 cfm (50 cfm – 37 cfm). For the other bathroom, the Initial Room Airflow Deficit is 50 cfm (50 – 0). For the kitchen it is zero, because 126 cfm exceeds the required 100 cfm. The excess capacity provided by the kitchen exhaust is not included.

Step 2 is to determine the Window Opening Credit for each room. The first bathroom and kitchen do not have openable windows, but the second bathroom does. The second bathroom receives a credit of 20 cfm. The final room airflow deficit for the first bathroom is 13 cfm, and for the second bathroom it is 30 cfm (50 cfm – 20 cfm).

Step 3 is to sum the final airflow deficit for each room and compute the required additional airflow. The total airflow deficit for the house is 43 cfm (13 cfm plus 30 cfm). The excess capacity provided by the kitchen exhaust is not included. The required additional airflow is equal to one-quarter of the total airflow deficit. This is 11 cfm (43 cfm divided by 4). The capacity of the whole-building ventilation system must be 11 cfm greater than the value determined from Section 4.1 in order to compensate for the local exhaust ventilation deficiencies.

Example A-E Determining the Required Whole-Building Ventilation Rate

Q: An existing 2,350 square foot home has three bedrooms, two bathrooms and a kitchen. One bathroom has an exhaust fan which has a measured airflow of 37 cfm, but no openable window. The other bathroom has no fan but does have an openable window. The kitchen has an exhaust fan with a measured airflow of 26 cfm and two openable windows. The bath and kitchen fans are not being replaced. A blower-door test and calculations performed in accordance with ASHRAE Standards 119 and 136 show an infiltration credit of 19 cfm. A new central exhaust-only ventilation fan will be installed. What is the required airflow for the new fan?

A: Calculating the required airflow is a multistep process as follows.

Step 1 is to determine the required ventilation rate from Section 4.1. Using Table 4.1a, we get 60 cfm (using Equation 4.1a we would get 58.5 cfm).

Step 2 (from Section A3.1) is to determine the Initial Room Airflow Deficit for each room. For the bathrooms, Table 5.1 requires an exhaust capacity of 50 cfm with 100 cfm required for the kitchen. The Initial Room Airflow Deficit is the required airflow from Table 5.1 minus the airflow rating of the existing fan. For the first bathroom, this value is 13 cfm (50 cfm – 37 cfm). For the other

bathroom, the Initial Room Airflow Deficit is 50 cfm (50 – 0). For the kitchen it is 74 cfm (100 – 26).

Step 3 (from Section A3.2) is to determine the Window Opening Credit for each room. The first bathroom does not have openable windows, but the second bathroom does and the kitchen has two. The second bathroom receives a credit of 20 cfm. The kitchen also receives a 20 cfm credit because only one window credit is available per room, regardless of the number of openable windows in the room. The final room airflow deficit for the first bathroom is 13 cfm, for the second bathroom it is 30 cfm (50 cfm – 20 cfm), and for the kitchen it is 54 cfm (74 – 20).

Step 4 (from Section A3.3) is to sum the final airflow deficits for each room and compute the required additional airflow (the additional whole-building ventilation required to make up for the deficits). The total airflow deficit for the house is 97 cfm (13 cfm + 30 cfm + 54 cfm). The required additional airflow is equal to one-quarter of the total airflow deficit. This is 25 cfm (97 cfm divided by 4, rounded up).

Step 5 is to add the required additional airflow to the required ventilation rate from Section 4.1. The total required ventilation rate is the now 85 cfm (60 cfm plus 25 cfm).

Step 6 is to determine the required fan capacity by deducting the infiltration credit, determined in Section 4.1.3 (19 cfm) from the total required ventilation rate. The required fan capacity is 66 cfm (85 cfm – 19 cfm).

Appendix B – Addenda Description Information

Appendix B simply describes the addenda to Standard 62.2-2007 that have been published and have become part of the standard. This User's Manual is based on Standard 62.2-2010, which includes Addenda 62.2a, b, c, d, e, f, g, h, i, j, k, m, n, o, p, q, r, and t to 62.2-2007. As ASHRAE continues to publish additional addenda, various requirements in the standard may change. Appendix B of future versions will list those addenda, and may indicate areas where this User's Manual is no longer correct due to those changes.